Current Topics in Microbiology 227 and Immunology

Editors

R.W. Compans, Atlanta/Georgia
M. Cooper, Birmingham/Alabama
J.M. Hogle, Boston/Massachusetts · Y. Ito, Kyoto
H. Koprowski, Philadelphia/Pennsylvania · F. Melchers, Basel
M. Oldstone, La Jolla/California · S. Olsnes, Oslo
M. Potter, Bethesda/Maryland · H. Saedler, Cologne
P.K. Vogt, La Jolla/California · H. Wagner, Munich

Springer
Berlin
Heidelberg
New York
Barcelona
Budapest
Hong Kong
London
Milan
Paris
Santa Clara
Singapore
Tokyo

Cyclin Dependent Kinase (CDK) Inhibitors

Edited by P.K. Vogt and S.I. Reed

With 15 Figures and 7 Tables

Professor Peter K. VOGT, Ph.D.
Division of Oncovirology
Department of Molecular and Experimental Medicine
The Scripps Research Institute
SBR7, Rm. SR311
10550 North Torrey Pines Road
La Jolla, CA 92037
USA

Professor Steven I. REED, Ph.D.
Department of Molecular Medicine, MB-7
The Scripps Research Institute
10550 North Torrey Pines Road
La Jolla, CA 92037
USA

Cover Illustration: p27 nullizygous mouse, with wild-type control for comparison. Gigantism is one of the phenotypes conferred by germline mutation of p27 (see Kiyokawa and Koff, this volume).

Cover Design: design & production GmbH, Heidelberg

ISSN 0070-217X
ISBN 3-540-63429-0 Springer-Verlag Berlin Heidelberg New York

This work is subject to copyright. All rights are reserved, whether the whole or part of the material is concerned, specifically the rights of translation, reprinting, reuse of illustrations, recitation, broadcasting, reproduction on microfilm or in any other way, and storage in data banks. Duplication of this publication or parts thereof is permitted only under the provisions of the German Copyright Law of September 9, 1965, in its current version, and permission for use must always be obtained from Springer-Verlag. Violations are liable for prosecution under the German Copyright Law.

© Springer-Verlag Berlin Heidelberg 1998
Library of Congress Catalog Card Number 15-12910
Printed in Germany

The use of general descriptive names, registered names, trademarks, etc. in this publication does not imply, even in the absence of a specific statement, that such names are exempt from the relevant protective laws and regulations and therefore free for general use.

Product liability: The publishers cannot guarantee the accuracy of any information about dosage and application contained in this book. In every individual case the user must check such information by consulting other relevant literature.

Typesetting: Scientific Publishing Services (P) Ltd, Madras
SPIN: 10629157   27/3020 – 5 4 3 2 1 0 – Printed on acid-free paper

Preface

More than 10 years ago, the discovery of cyclin-dependent kinases (Cdks) ushered in a new era in the understanding of cell proliferation and its control. Not only were both of the known cell cycle transitions, from G1 to S phase and G2 to M phase, found to be dependent on these protein kinases, but the regulatory assumption intrinsic to cyclin-dependent kinases, a stable inactive catalytic subunit (the Cdk) and an unstable requisite positive regulatory activating subunit (the cyclin), led to a simple model for cell cycle control. Modulation of cyclin accumulation, and thereby Cdk activation, was proposed to be the overarching principle governing the passage through cell cycle phases. Another reality to emerge from the discovery of Cdks was the exceptional degree of evolutionary conservation maintained in the machinery and organization of proliferation control. Not only were Cdks shown to be structurally conserved between yeast and man, but mammalian Cdks could substitute functionally for the endogenous enzymes in a yeast cell. The problem of cell cycle control was thought to have been virtually solved.

The ensuing years have provided a much more complex view of cell cycle control and the role and regulation of Cdks. The uncritical enthusiasm with which many of the ideas were embraced has required tempering. For example, although Cdks appear to be highly conserved phylogenetically, cyclins are much less so. Classes of cyclins that are critical for cell cycle control in yeast apparently do not exist in mammalian cells and vice versa. More significantly, modulation of cyclin accumulation is not the all-encompassing regulatory principle initially envisioned. Certainly, in some specialized cell cycles, such as the cleavage divisions of early amphibian embryos, the accumulation and degradation of cyclins appears to be the driving force behind cell cycle progression. However, in most situations, whereas cyclin accumulation is essential, it is not rate-limiting for progression from one cell cycle phase to the next. Intense investigation into the structure and function of Cdks since their discovery has revealed several additional regulatory modes superimposed on the basic Cdk-cyclin motif. Regulatory phosphorylation of Cdks appears to be a conserved and pervasive mechanism for many cell types. This volume, however, deals with a second ubiquitous mode of Cdk regulation, that by Cdk regulation, that by Cdk inhibitory proteins.

Although Cdk inhibitory proteins, known also as CKIs, have emerged relatively recently (within the past 3–5 years) in the arena of cell cycle control, they now constitute one of the principal foci of intellectual and experimental activity in the cell cycle field. Interest in CKIs has been further stimulated by the demonstration that at least one them is an important tumor suppressor, and that mutations in the corresponding gene are associated with human malignancy. Surprisingly, there have been relatively few comprehensive review articles and, as far as we know, no complete volumes dedicated to this subject. We felt that, with the volume of information accumulated and the importance of the subject matter to a broad spectrum of biomedical scientists, the time was right for assembling a comprehensive collection of reviews on CKIs by experts in the relevant subdisciplines. The first chapter (Mendenhall) discusses the roles of CKIs where they were first discovered and are best understood, in yeast. Even though the yeast CKIs are not close structural relatives of mammalian CKIs, the precision of genetic analysis in yeast allows insights that are likely to be useful in conceptualizing the roles and regulation of mammalian CKIs. The two next chapters comprise general reviews describing the properties of members of the two known families of mammalian CKIs. Members of the Cip/Kip family (Hengst and Reed) are general inhibitors targeted to a broad spectrum of Cdks, whereas members of the INK4 family (Carnero and Hannon) are specific for a specialized Cdk subfamily composed of Cdk4 and Cdk6. The ensuing chapters deal with more specific aspects of CKI biology. Chellappan et al. discuss the role of CKIs and concomitant Cdk regulation in the context of cellular differentiation and organismic development. Kiyokawa and Koff approach this same topic specifically through the technique of gene disruption in mice. These studies continue to be the most revealing concerning CKI function in vivo and, at the same time, to provide the most unexpected results. The next two chapters deal with the issues of genome integrity and malignancy. Although many CKIs have clearly demonstrable roles in controlling cell proliferation, few appear to be implicated in malignancy. Kamb summarizes the current data linking loss of CKI function with human malignancy. El-Deiry addresses the special relationship of the CKI p21 to the tumor suppressor p53, strongly implicated in maintenance of genome integrity and prevention of malignancy. Finally, the last chapter by Walker looks at Cdk inhibition from an entirely different perspective: as a potential therapeutic target. If endogenous protein inhibitors of Cdks can have potent effects on cell proliferation and survival, it follows that small molecule Cdk inhibitors may be unique tools to effect these processes in a clinical context.

The biology of CKIs is a young field still in a high state of flux and accommodates many conflicting opinions. The fact that several of the chapters deal with overlapping subjects but from different experimental and personal perspectives ensures that most, if not all, of the current views are represented. It is hoped that this book will serve as a dispassionate source of information for both experts and non-experts. We would, in particular, like to thank the authors for their efforts in preparing up-to-date reviews in the context of an extremely rapidly moving field and a large volume of sometimes conflicting data. It is our sincere hope that this work will help stimulate interest and investigation in the cell cycle field and ultimately, by so doing, advance the causes of human knowledge and human health.

La Jolla, California
November, 1997

Steven I. Reed
Peter K. Vogt

List of Contents

M.D. Mendenhall
Cyclin-Dependent Kinase Inhibitors of
Saccharomyces cerevisiae and *Schizosaccharomyces pombe* 1

L. Hengst and S.I. Reed
Inhibitors of the Cip/Kip Family 25

A. Carnero and G.J. Hannon
The INK4 Family of CDK Inhibitors 43

S.P. Chellappan, A. Giordano, and P.B. Fisher
Role of Cyclin-Dependent Kinases and
Their Inhibitors in Cellular Differentiation
and Development 57

H. Kiyokawa and A. Koff
Roles of Cyclin-Dependent Kinase Inhibitors:
Lessons from Knockout Mice 105

W.S. El-Deiry
p21/p53, Cellular Growth Control and Genomic
Integrity 121

A. Kamb
Cyclin-Dependent Kinase Inhibitors
and Human Cancer 139

D.H. Walker
Small-Molecule Inhibitors of Cyclin-Dependent Kinases:
Molecular Tools and Potential Therapeutics 149

Subject Index 167

List of Contributors

(Their addresses can be found at the beginning of their respective chapters.)

CARNERO, A.   43

CHELLAPPAN, S.P.   57

EL-DEIRY, W.S.   121

FISHER, P.B.   57

GIORDANO, A.   57

HANNON, G.J.   43

HENGST, L.   25

KAMB, A.   139

KIYOKAWA, H.   105

KOFF, A.   105

MENDENHALL, M.D.   1

REED, S.I.   25

WALKER, D.H.   149

Cyclin-Dependent Kinase Inhibitors of *Saccharomyces cerevisiae* and *Schizosaccharomyces pombe*

M.D. MENDENHALL[1,2]

1	Introduction	1
2	Pho81	2
2.1	Pho81 Biochemistry	2
2.2	*PHO81* Genetics	3
3	SIC1	5
3.1	General Properties	6
3.2	Sic1 and the G_2/M Phase Transition	7
3.3	Sic1 and the G_1/S Phase Transition	8
3.4	Regulation of Sic1 Levels	8
3.5	Consequences of a *sic1* Deletion	9
4	FAR1	10
4.1	Regulation of CDK Activity by Far1	11
4.2	Regulation of Far1 Levels	13
4.3	Regulation of Far1 Activity	14
4.4	A Non-CKI Role for Far1	15
4.5	*FAR1* and the Cell Cycle	16
5	RUM1	17
5.1	Rum1 as an Inhibitor of Mitosis	17
5.2	Rum1 as a Promoter of S Phase	18
5.3	Rum1 and the Establishment of Pre-Start G_1	19
6	Conclusions	20
	References	20

1 Introduction

The yeasts *Saccharomyces cerevisiae* and *Schizosaccharomyces pombe* have been favored organisms for the study of the basic biology, genetics, and biochemistry of the mitotic cycle. Much of what we understand about eukaryotic cell division derives from studies in these organisms. Cyclin-dependent protein kinase (CDK) inhibitors (CKI), the focus of this volume, were first described in yeast, and yeast is still the best system for dissecting out the complex in vivo relationships between the CKI and other cell cycle components. This article summarizes the current state of

[1] Department of Biochemistry, University of Kentucky, Lexington, KY 40536-0096, USA
[2] L.P. Markey Cancer Center, University of Kentucky, 800 Rose St., Lexington, KY 50536-0096, USA

the literature on four CKI – Pho85, Sic1, Far1, and Rum1 (Table 1) – with a particular emphasis on placing their activities in the perspective of larger cellular processes.

Despite the common biochemical activity linking these four proteins, each goes about its business in a unique way. Pho85 and Far1 link signal transduction pathways to specific cellular responses, i.e., nutritional limitation to gene transcription and pheromone detection to cell cycle arrest, respectively. In addition to being a CKI, both of these proteins have non-CKI functions. Pho85 works directly with a transcriptional activator and a cyclin (without a CDK) to control gene transcription. Far1 has a role in determining the cell surface site at which polarized growth will occur. Sic1 and Rum1 are intrinsic cell division cycle regulators responsible for ensuring the proper coordination of critical mitotic events. Both Sic1 and Rum1 are G_1 phase-expressed proteins, but have interactions with multiple CDK–cyclin partners (interactions that are not always inhibitory) that affect events in all parts of the cell cycle.

2 Pho81

Saccharomyces cerevisiae requires inorganic phosphate (P_i) for growth and division. When the extracellular concentration of P_i becomes limiting, genes encoding secreted phosphatases are derepressed. The phosphatases cleave P_i from organic phosphates, making the P_i available to the cell (for a review, see OSHIMA 1982). Pho81 encodes a CKI involved in the transcriptional regulation of the secreted phosphatases.

2.1 Pho81 Biochemistry

Pho81 can claim to be the first CKI, as it was identified as a genetic entity involved in the regulation of phosphate metabolism in 1973 (TOH-E et al. 1973), although the realization that it is a CKI was much more recent (K.R. SCHNEIDER et al. 1994). With 1117 amino acid residues and a molecular mass of 134 kDa, Pho81 (GenBank

Table 1. The yeast CKI and the CDK and cyclins they inhibit

CKI	CDK	Cyclin
Pho81	Pho85	Pho80
Far1	Cdc28	Cln1, Cln2
Sic1	Cdc28	Clb2, Clb5
Rum1	Cdc2	Cdc13, Cig2

CDK, cyclin-dependent kinase; CKI, CDK inhibitor.

X87941) is the largest of the known CKI but, with six ankyrin repeats (residues 424-656), it has homology to the smallest class of mammalian CKI (OGAWA et al. 1993; K.R. SCHNEIDER et al. 1994). Pho81 is an effective inhibitor of the Pho85/Pho80 CDK/cyclin complex, with a reported IC_{50} of 1 nM (K.R. SCHNEIDER et al. 1994), a value comparable to that obtained with CKI in other systems (MENDENHALL 1993; HARPER et al. 1995). The effect of Pho81 on other CDK/cyclin complexes has not been reported.

Pho81 is not an ATPase, nor does it have phosphatase or protease activity on Pho85–Pho80 substrates (K.R. SCHNEIDER et al. 1994), but acts primarily through its physical interaction with Pho85–Pho80 complexes. Surprisingly, given its sequence homology with $p15^{INK4b}$ and $p16^{INK4a}$, which bind to the protein kinase domain of their CDK–cyclin target (HALL et al. 1995), Pho81 recognizes only the Pho80 cyclin component. In both co-immunoprecipitation and two-hybrid assays, Pho81 forms complexes with Pho80 in cells deleted for Pho85 and forms complexes with the Pho85 protein kinase only when Pho80 is also present (HIRST et al. 1994; K.R. SCHNEIDER et al. 1994).

The ankyrin repeat domain possesses at least some of the Pho81 CKI activity. When purified from a bacterial expression system, Pho81 residues 400–720 are able to inhibit Pho85–Pho80, but with a 60-fold higher IC_{50} than the whole protein purified from yeast (K.R. SCHNEIDER et al. 1994). In vivo, expression of only the fifth and sixth ankyrin repeats (residues 584-724) confers substantial, though again incomplete, activity (OGAWA et al. 1995). Deletion analysis of Pho81 has indicated that the amino terminal domain has a negative effect on Pho81 inhibitory activity, while the carboxy-terminal domain antagonizes this negative effect even when supplied in *trans* (OGAWA et al. 1995).

2.2 PHO81 Genetics

The early identification of *PHO81* came in studies of the mechanisms governing the response of *S. cerevisiae* to limiting extracellular concentrations of P_i. Under these conditions, genes encoding a high-affinity P_i transporter (*PHO84*), an alkaline phosphatase (*PHO8*), and three acid phosphatase isozymes (*PHO5, PHO10*, and *PHO11*) are expressed (TOH-E et al. 1973, 1976; BUN-YA et al. 1991; OGAWA et al. 1995). Addition of P_i to the growth medium represses transcription of all these genes, a response controlled by two positive factors, *PHO81* and *PHO4*, and two negative factors, *PHO80* and *PHO85*. The genetic properties of the regulatory genes and the epistasis relationships between them provided the basis for the model proposed by OSHIMA (1982) and shown in Fig. 1.

DNA sequence analyses of these genes and the subsequent biochemical studies prompted by the sequence data revealed that *PHO85* encoded a CDK (UESONO et al. 1987; TOH-E et al. 1988) and that *PHO80* encoded its activating cyclin (KAFFMAN et al. 1994). Pho4 is a transcription factor that binds the Pho5 promoter through a basic helix–loop–helix motif (VOGEL et al. 1989; F. FISHER et al. 1991) and is also the favored substrate for the Pho85–Pho80 protein kinase (KAFFMAN

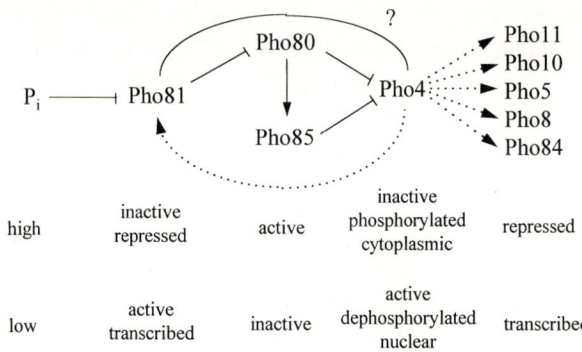

Fig. 1. Regulatory circuitry centering on Pho81. *Arrows* indicate a positive interaction, —| indicates a negative interaction, *solid lines* indicate regulatory events at the protein level, and *dotted lines* indicate regulatory events at the transcriptional level. The depicted interactions are not necessarily direct. P_i inorganic phosphate. Pho81 is a cyclin-dependent kinase (CDK) inhibitor (CKI); Pho80 is a cyclin; Pho85 is a CDK; Pho4 is a transcriptional activator; Pho5, Pho10, and Pho 11 are acid phosphatase isozymes; Pho8 is alkaline phosphatase; and Pho84 is a high-affinity P_i transporter. At the *bottom*, the states of the various components in either high or low [P_i] are indicated. Experimental observations are discussed in the text and references cited therein

et al. 1994). The phosphorylation of Pho4 by Pho85–Pho80 is presumably an important aspect of the regulation of Pho4 transcription activation activity, but it is still a matter of debate as to just how important it is (HIRST et al. 1994; O'NEILL et al. 1996).

Genetic experiments identified Pho81 as the primary negative regulator of the Pho85–Pho80 complex. Strains carrying recessive mutations in *pho81* do not express *PHO5* under low- or high-P_i conditions, but strains carrying *pho81 pho80* and *pho81 pho85* double mutations constitutively express *PHO5* in a P_i-independent manner similar to strains carrying *pho80* or *pho85* single mutations (TOH-E et al. 1973; UEDA et al. 1975). These results indicate that Pho81 is a positive regulator of *PHO5* transcription which acts by negatively regulating Pho80 or Pho85 activity and that Pho81 activity is, in turn, negatively regulated by P_i.

For the most part, the predictions made by this model have been confirmed by the identification of Pho81 as a CKI for Pho80–Pho85 complexes, but it is not yet understood how P_i regulates this system. Pho85–Pho80 activity from cells grown in high P_i medium was five times that of cells grown in limiting P_i. Cells lacking *pho81* had high Pho85–Pho80 activity regardless of whether they were grown in high- or in low-P_i conditions, strengthening the case that Pho81 is a mediator of P_i repression (K.R. SCHNEIDER et al. 1994). Artificial overexpression of Pho81 led to a reduction in measurable Pho85–Pho80 activity. Since *PHO81*, like the repressible phosphatase genes, is under *PHO4* transcriptional control (YOSHIDA et al. 1989), a model for *PHO* gene control involving a positive feedback loop that would increase Pho81 protein levels in response to low-P_i conditions seemed reasonable.

Numerous observations have accumulated to make this simple model untenable. LEMIRE et al. (1985) found that derepression of *PHO5* occurs when protein synthesis is inhibited by cycloheximide, suggesting that increased Pho81 transcription is not important. In agreement with this observation, K.R. SCHNEIDER et al. (1994) found that the amount of Pho81 present in Pho85–Pho80 complexes is unchanged whether cells are grown in low or high phosphate media. This suggested that the activity of Pho81 might be altered by the different growth conditions, but no intrinsic differences in inhibitory activity were found when Pho81 was prepared from cells grown in low- versus high-P_i conditions (K.R. SCHNEIDER et al. 1994). Pho81 activity could be regulated by labile modifications, weak protein–protein interactions, or alterations in cellular localization; at present, however it is not clear whether and, if so, how P_i alters Pho81.

Another series of results indicated that inhibition of Pho85–Pho80 kinase activity may be a relatively minor role of Pho81. Deletion of *PHO85* results in a constitutive level of acid phosphatase transcription that is only 15% of the level obtained when *PHO80* is deleted, indicating that the cyclin component is more important than the kinase component for maintaining repression of *PHO5* (LEMIRE et al. 1985). Consistent with this, a mutant allele of *PHO4* that lacks all of the Pho85–Pho80-dependent phosphorylation sites and, therefore, should be independent of Pho85–Pho80 activity expresses only 15% of the fully induced levels of acid phosphatase activity and is still responsive to extracellular P_i levels (O'NEILL et al. 1996). Furthermore, overexpression of *PHO80* established full repression of acid phosphatase in a strain that was deleted for *PHO85*, indicating that the kinase component was not necessary for repression of Pho4 activity (MADDEN et al. 1990). In this latter case, it could be argued that high levels of Pho80 could activate another CDK that could then phosphorylate Pho4, but the relative unimportance of Pho85–Pho80 phosphorylation in the control of Pho4 activity seems clear.

A possible alternative activity for Pho81 was indicated by the two-hybrid results obtained by HIRST et al. (1994). They reported evidence for a physical interaction between Pho81 and residues 1-73 of Pho4. This interaction did not require the presence of Pho80, but Pho80 did enhance the Pho81–Pho4 interaction signal. Pho80–Pho4, Pho81–Pho4, and Pho81–Pho80 interactions were seen in high-P_i (repressing) conditions, but only Pho81–Pho80 interactions were detected in low-P_i (inducing) conditions. Unfortunately, these two-hybrid results could not be confirmed by co-immunoprecipitation of Pho4 and Pho81. Nevertheless, a role for Pho80 as a direct inhibitor (not requiring the kinase subunit) of the Pho4 transcription factor with Pho81 acting to antagonize Pho80 is suggested.

3 Sic1

Major events in the *S. cerevisiae* cell cycle are controlled by the activity of the Cdc28 protein kinase in association with at least nine different cyclins. With the exception of *CLN3*, the cyclin genes come in homologous pairs, and each pair has a

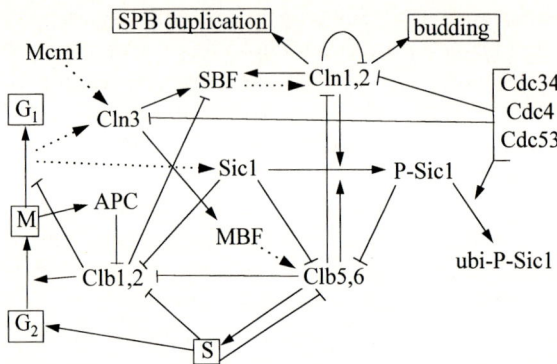

Fig. 2. Regulatory circuitry centering on Sic1. Conventions are as for Fig. 1. *SPB*, spindle pole body; *APC*, anaphase-promoting complex. Cln1, Cln2, Cln3, Clb1, Clb2, Clb5, and Clb6 are cyclins; Sic1 is a cyclin-dependent kinase (CDK) inhibitor (CKI; P-Sic1 is the phosphorylated form and ubi-P-Sic1 is the ubiquitinated form); SBF and MBF are transcription factors; APC and a putative complex consisting of Cdc34/Cdc4/Cdc53 are ubiquitinating factors active at anaphase and at Start, respectively. The cyclins are active when combined with the Cdc28 CDK, which is not shown. Note the overlap with Fig. 3. Experimental observations are discussed in the text and references cited therein

characteristic cell cycle interval during which it is transcribed and in which it activates Cdc28 to carry out specific functions. *CLN3* has the primary responsibility of determining whether the cell has grown large enough for a new mitotic cycle and whether the extracellular environment will support a new mitotic cycle. *CLN1*, *CLN2*, *CLB5*, and *CLB6* appear in late G_1 phase. *CLN1* and *CLN2* have a primary role in bud initiation and spindle pole duplication. *CLB5* and *CLB6* initiate S phase and reshape the bud growth. *CLB3* and *CLB4* appear next and have a role in late S-phase events. *CLB1* and *CLB2* appear last and are responsible for initiating mitotic chromosome segregation (Lew and Reed 1993; Schwob and Nasmyth 1993; Dirick et al. 1995; Breeden 1996). An elaborate network of regulatory relationships among these genes and their gene products ensures that these events proceed in an orderly process (see Fig. 2). Sic1 is a CKI that ensures that the Cdc28–Cln1,2 complexes are activated before the Cdc28–Clb5,6 complexes.

3.1 General Properties

Sic1 (GenBank U01300) is the first CKI to have its inhibitory activity demonstrated biochemically (Mendenhall 1993). It is a 32-kDa protein composed of 284 amino acids and was originally discovered as a Cdc28 substrate that co-immunoprecipitated with Cdc28 (Reed et al. 1985). At that time it was called p40 due to its apparent molecular weight on (sodium dodecyl sulfate polyacrylamide gel electrophoresis (SDS-PAGE). The name *SIC1* is derived from the observation that p40 is both a *s*ubstrate and an *i*nhibitor of *C*dc28 (Nugroho and Mendenhall 1994). Using its ability to be phosphorylated by Cdc28–Clb2 complexes as an assay, the Sic1 protein was purified (Mendenhall 1993) and its gene cloned using oligonucleotides predicted from the protein sequence (NUGROHO and MENDENHALL 1994). Whether purified from yeast or from a bacterial expression system, Sic1 is a potent inhibitor of Cdc28–Clb complexes with a K_i of less than 1 nM (MENDENHALL 1993; SCHWOB et al. 1994; MENDENHALL et al. 1995). It is not an inhibitor of Cdc28–Cln complexes (SCHWOB et al. 1994). It was also shown not to be a histone H_1 phosphatase or protease or an ATPase (MENDENHALL 1993).

Sic1 is only found in G_1-phase cells (MENDENHALL et al. 1987; DONOVAN et al. 1994). This phase specificity is a reflection of periodic *SIC1* transcription, which rises in late mitosis and declines during G_1 phase (DONOVAN et al. 1994; SCHWOB et al. 1994). Periodicity at the protein level also implies that Sic1 turnover is rapid, something which, as it turns out, is under separate cell cycle control (see below).

3.2 G_2/M-Phase Transition

The G_1-phase specificity of Sic1 expression and its specificity for Clb cyclins suggested that it would function as an inhibitor of Cdc28–Clb1 and Cdc28–Clb2 complexes in the completion of M phase or that it would act to delay S-phase initiation by inhibiting Cdc28–Clb5 and Cdc28–Clb6 complexes. Some of the phenotypes of the *sic1* deletion are consistent with a role in mitosis; *sic1*-Δ cultures accumulate large numbers of cells arrested in medial nuclear division, although the majority of the cells grow and divide relatively normally. Flow cytometry indicates that G_1-phase cells are virtually absent, consistent with a mitotic block (NUGROHO and MENDENHALL 1994).

In support of this interpretation, DONOVAN et al. (1994) independently isolated *SIC1* as *SDB25*, a high copy suppressor of *dbf2*[ts] mutations. Dbf2 is a protein kinase with complex genetics (JOHNSTON et al. 1990; TOYN et al. 1991) that indicate a probable, but as yet undefined role in late nuclear division (JOHNSTON et al. 1990; KITADA et al. 1993; TOYN and JOHNSTON 1994; MORISHITA et al. 1995). It is periodically expressed in the cell cycle 15 min prior to the expression of *SIC1* (JOHNSTON et al. 1990; DONOVAN et al. 1994). In addition to *SIC1*, other high copy suppressors of *dbf2*[ts] include the protein kinases *CDC5* and *CDC15*. Cdc5 and its homologues from *Drosophila* (polo) and human (Plk1) have been implicated in the anaphase to telophase transition (PRINGLE and HARTWELL 1981; CLAY et al. 1993; FENTON and GLOVER 1993). Perhaps more enlightening, Cdc15 has been implicated in the destruction of the Clb2 cyclin at anaphase (SURANA et al. 1993). The common thread among all these *dbf2*[ts] suppressors is that, directly or indirectly, they reduce the activity of the Cdc28–Clb2 kinase. As far as *SIC1* is concerned, this activity might be artifactual, since the high copy suppression may have removed the normal temporal controls over *SIC1* expression, allowing it to carry out an activity that it normally would not have carried out. Arguing against this interpretation is the observation that deletions of *sic1* and *dbf2* are synthetically lethal (despite the recessive temperature-sensitive lethality of *dbf2*[ts] mutations, deletions of *dbf2* are viable) (DONOVAN et al. 1994).

3.3 G₁/S-Phase Transition

Mutations in *dbf2* also display a delay in S-phase initiation (JOHNSTON et al. 1990) that may be part of a newly described checkpoint acting in late G_1 phase to prevent premature mitosis (TOYN et al. 1995). The interaction between *SIC1* and *DBF2* described in the previous paragraph may therefore result from a role for Sic1 in late G_1 phase, a role for which there is abundant support.

In their studies of the role of the Cdc28–Clb5 kinase, SCHWOB et al. (1994) found that, although Clb5 and Cln2 were transcribed at the same time, Cdc28–Clb5 protein kinase activity was delayed relative to that of Cdc28–Cln2. This delay was abolished in a *sic1-Δ* strain, indicating that Sic1 was the cause of the delay. Furthermore, by deleting all six *CLB* genes, they were able to show that *CLB* gene expression was required for S-phase initiation. In normal cycling cells, *CLB5* and *CLB6* are the only *CLB* genes expressed at the G_1/S-phase border (EPSTEIN and CROSS 1992; KÜHNE and LINDER 1993; SCHWOB and NASMYTH 1993). These results indicate a role for Sic1 in the delay of S-phase initiation. Consistent with this, *sic1-Δ* strains replicate DNA much earlier than $SIC1^+$ strains (SCHWOB et al. 1994; B.L. SCHNEIDER et al. 1996).

3.4 Regulation of Sic1 Levels

For S phase to proceed, the inhibition of Cdc28–Clb activity conferred by Sic1 would need to be reversed. Since *SIC1* is periodically transcribed and encodes an unstable protein (MENDENHALL 1993; DONOVAN et al. 1994; SCHWOB et al. 1994), it seemed possible that transcriptional controls are all that are needed. However, constitutively expressed *SIC1* was unable to cause permanent cell cycle arrest (NUGROHO and MENDENHALL 1994), suggesting the existence of an additional mechanism affecting Sic1 inhibitory ability. An important clue to this mechanism came from the phenotype of the strain carrying the deletions in all six *CLB* genes. Despite being unable to initiate DNA synthesis, the Δ*clb* strain budded repeatedly, producing mononucleate cells with multiple attached buds (SCHWOB et al. 1994). Cells with this morphology were occasionally observed in strains constitutively overproducing *SIC1* (NUGROHO and MENDENHALL 1994), indicating a link between high Sic1 activity and low Clb activity.

This multibudded morphology is also displayed by $cdc34^{ts}$ mutations (PRINGLE and HARTWELL 1981). *CDC34* encodes a ubiquitin transferase of the E2 class (GOEBL et al. 1988) and is involved in the degradation of many G_1 phase-limited proteins (DESHAIES et al. 1995; YAGLOM et al. 1995; LANKER et al. 1996). If Cdc34 were required for the degradation of Sic1, an event needed for the release of Cdc28–Clb kinase activity, the multibudded morphology of the Δ*clb* and $cdc34^{ts}$ strains would have a common basis. That this was true was dramatically demonstrated by showing that a $cdc34^{ts}$ *sic1-Δ* strain no longer had the multibudded morphology and was able to replicate its DNA (SCHWOB et al. 1994; MENDENHALL et al. 1995). Furthermore, the Sic1 protein became hyperstabilized in a $cdc34^{ts}$ background at the restrictive temperature (SCHWOB et al. 1994). These results imply not only that Sic1 degradation is mediated by Cdc34-dependent ubiquitination, but that Sic1 is the only substrate whose degradation is required for S-phase entry. Two other genes, *CDC4* and *CDC53*, have the same multibudded morphology, which is similarly suppressed by the *sic1*-Δ mutation, indicating a role for these genes in this process as well.

Other instances of Cdc34-mediated ubiquitination seem to require prior phosphorylation by Cdc28–Cln complexes (DESHAIES et al. 1995; YAGLOM et al. 1995; LANKER et al. 1996). Sic1 does not seem to be an exception. Sic1 is phosphorylated by Cdc28–Cln2 complexes in vitro (SCHWOB et al. 1994) and is known to be a phosphoprotein in vivo (DONOVAN et al. 1994). Furthermore, using a system which can be switched from a Cln^+ $Cdc34^-$ state to a Cln^- $Cdc34^+$ state, B.L. SCHNEIDER et al. (1996) showed that Sic1 accumulates in a hyperphosphorylated state when Cln activity is high and Cdc34 activity low and rapidly becomes hypophosphorylated and stable when Cln activity is low and Cdc34 activity high. Incidentally, reciprocal shift experiments with this strain indicated that the Cln and Cdc34 activities were interdependent, which was not expected on the basis of earlier experiments done using *cdc28* and *cdc4* mutants (HEREFORD and HARTWELL 1974).

Deletion of *sic1* surprisingly restored the ability of $cdc34^{ts}$ strains to replicate their DNA, indicating that the sole cause of the DNA replication defect of *cdc34* mutant strains was their inability to degrade Sic1. Even more surprising is the observation that deletion of *sic1* restores viability to a strain lacking all three G_1-phase cyclins. EPSTEIN and CROSS (1994, personal communication) found that all 12 mutations that they had isolated allowing a *cln1*-Δ *cln2*-Δ *cln3*-Δ triple mutation to survive mapped to a single locus that was tightly linked to *SIC1*. The phenotypic properties conferred by the mutation, which they named *BYC1*, resembled those of a *sic1*-Δ strain. A directed deletion of *SIC1* in a *cln1*-Δ *cln2*-Δ *cln3*-Δ background confirmed the result (B.L. SCHNEIDER et al. 1996; TYERS 1996). This finding indicates that the only *essential* function of the *CLN* genes is to participate in the degradation of Sic1 protein. This is not to say that this is the only function. The *cln1*-Δ *cln2*-Δ *cln3*-Δ *sic1*-Δ strain has a highly abnormal morphology, consistent with an important role for the *CLN* genes in the proper initiation of the nascent bud (LEW and REED 1993).

3.5 Consequences of a *sic1* Deletion

The preferential inhibition by Sic1 of Cdc28–Clb complexes relative to Cdc28–Cln complexes and the linkage of Sic1 degradation to Cdc28–Cln-dependent phosphorylation creates a condition in which Sic1 delays the activation of Cdc28–Clb complexes relative to Cdc28–Cln complexes. The consequences of the loss of this delay are quite severe. *sic1*-Δ strains are viable, but have a plating efficiency that is only 60% of wild-type cells (EPSTEIN and CROSS 1994; NUGROHO and MENDENHALL 1994). The subpopulation that was incapable of further cell division arrested with large buds and an undivided G_2-phase nucleus that had migrated to the mother–daughter junction, but had not separated its chromatin (medial nuclear division; see PRINGLE and HARTWELL 1981). One possible interpretation of this phenotype consistent with the role for Sic1 in G_2/M phase (see above) was that these cells required Sic1 function to complete mitosis. An alternative hypothesis, consistent with the role in G_1/S phase just outlined is that premature activation of Cdc28–Clb complexes in G_1 phase leads to the frequent occurrence of a later cellular condition that cannot be resolved, resulting in permanent cell cycle arrest. For example, it may be necessary to complete events associated with nascent bud formation or with spindle pole body duplication (events that are initiated by Cdc28–Cln complexes) before DNA synthesis is initiated in order to ensure faithful DNA replication and chromosome separation. Alternatively, early activation of Cdc28–Clb6, which inhibits the activity of Cdc28–Cln complexes (BASCO et al. 1995), may prevent Cdc28–Cln complexes from carrying out important fidelity-ensuring events in the late G_1 phase.

One curious feature of the low plating efficiency seen in the $sic1$-Δ strain that deserves mention is that the inviable subpopulation is restricted to daughter cells, cells that had not previously produced a bud (NUGROHO and MENDENHALL 1994). No satisfactory explanation for this phenotype has been reported, but it suggests that $SIC1$ may have a function in setting up or monitoring the polarity of a critical cell cycle event. Another consequence of more general interest is the finding that $sic1$-Δ cells have highly elevated rates of chromosome breakage and loss. These chromosomal effects, usually associated with events in G_2/M phase, are apparently derived from the mistiming of events late in G_1 phase, although a direct role for $SIC1$ in late mitosis has not yet been ruled out. This connection between the timing of a major cell cycle event and later downstream effects on chromosomal integrity has important implications, because many of the events leading to tumorigenesis involve chromosomal rearrangements and because defects in cell cycle genes are increasingly implicated in tumorigenesis.

4 Far1

Haploid *S. cerevisiae* cells exist in two mating types, *MATa* and *MAT*α. When in contact, two cells of opposite mating types will fuse to form diploids in an elaborately controlled process (Fig. 3; for recent reviews, see BARDWELL et al. 1994; HERSKOWITZ 1995). The process is initiated by the exchange of peptide pheromones a-factor and α-factor, which bind to cell surface receptors on the opposite partner and activate a signal transduction pathway leading ultimately to diploid formation. An intermediate step in the process requires that the cell cycles of the two mating partners be synchronized. This is done by arresting cell division late in G_1 phase at Start, the yeast equivalent of the mammalian restriction point, by inactivating the Cdc28–Cln complexes. Far1 is the CKI responsible for this inactivation.

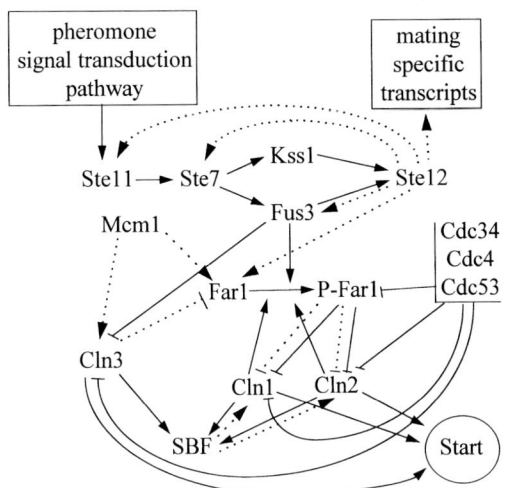

Fig. 3. Regulatory circuitry centering on Far1. Real, surmised, and supposed interactions among components of the mating pheromone signal transduction pathway. Conventions are as for Fig. 1. Ste11, Ste7, Kss1, and Fus3 are protein kinases; Ste12, Mcm1, and SBF are transcription factors; Cln1, Cln2, and Cln3 are cyclins; Far1 is a cyclin-dependent kinase (CDK) inhibitor (CKI; P-Far1 is a phosphorylated form); and Cdc34, Cdc4, and Cdc53 form a ubiquitin-transferase complex. The cyclins are active when combined with the Cdc28 CDK, which is not shown. Note the overlap with Fig. 2

4.1 Regulation of CDK Activity

Far1 is the first CKI to be genetically identified as a CKI (CHANG and HERSKOWITZ 1990). The gene encoding Far1 (GenBank Z49432) was originally isolated in a screen for mutations that would allow *S. cerevisiae* cells to grow in the presence of mating pheromone, but would not affect the mating pheromone-induced transcriptional response. The acronym FAR1 stands for *f*actor *a*rrest *r*esistant. Like Pho81, Far1 is a large CKI. It is composed of 830 amino acids and has a molecular mass of 95 000 kDa. Originally reported to be 780 amino acids in length (CHANG and HERSKOWITZ 1990), the DNA sequence was recently corrected (MCKINNEY and CROSS 1995) and 50 residues were added to the amino terminus of the open reading frame. The amino acid positions discussed in the early literature have been corrected in this review to fit the new data. The minimal region of Far1 necessary for CKI activity has not been defined systematically, but it lies in the amino terminal third and does not require the first 50 amino acids (CHANG and HERSKOWITZ 1990, 1992; ELION et al. 1993; PETER et al. 1993; MCKINNEY and CROSS 1995; VALTZ et al. 1995). Deletion of the amino terminus may affect Far1 specificity (MCKINNEY and CROSS 1995).

The isolation of *FAR1* as a gene required for mating pheromone-mediated arrest at Start immediately focused attention on Far1 interactions with the G_1-phase cyclins – Cln1, Cln2, and Cln3 – and the CDK that they activate, Cdc28. Far1 associated with Cln1, Cln2, and Cdc28 in co-immunoprecipitation experiments from yeast cell extracts (PETER et al. 1993; TYERS and FUTCHER 1993). Far1–Cdc28–Cln association was enhanced by prior treatment of the cells with mating pheromone. The association also required the activity of a mitogen-activated protein (MAP) kinase homologue called Fus3 known to be required for the mating pheromone response. Through the use of Far1 mutations and by varying mating pheromone dosage, it was shown that the ability of Far1 to associate with Cdc28–Cln2 correlated with the ability of Far1 to cause cell cycle arrest (PETER et al. 1993; TYERS and FUTCHER 1993). Far1 also associates with Cdc28–Cln3 in a mating pheromone dependent fashion (TYERS and FUTCHER 1993). The kinetics of this association are quite different, however. Far1–Cdc28–Cln3 complexes are only observed after an hour of mating pheromone treatment, whereas Far1–Cdc28–Cln2 complexes are seen in less than 5 min. The function of Far1–Cdc28–Cln3 association is not understood. It has been speculated that it could be involved in either the maintenance of or the recovery from mating pheromone arrest.

As Cdc28–Cln activity is normally required for passage through Start, it was proposed that Far1 is an inhibitor of Cdc28–Cln activity (CHANG and HERSKOWITZ 1990). This supposition was confirmed by PETER and HERSKOWITZ (1994), who showed that hexahistidine-tagged Far1 purified from mating pheromone-treated yeast and GST-tagged Far1-22 (from a constitutively active, mutant allele of $FAR1$) purified from bacteria was able to inhibit Cdc28–Cln2 protein kinase activity in vitro. Furthermore, Cdc28–Cln2 immunoprecipitated from cells treated with mating pheromone had only one fifth to one third of the activity of exponentially growing control cells. Stringent washing of the immunoprecipitates removed associated Far1 protein, restoring full protein kinase activity, and immunoprecipitations from cells lacking the $FAR1$ gene ($\Delta far1$) had uninhibited levels of Cdc28–Cln2 activity. In these experiments, Far1 did not affect the levels of Cln2 protein, the association of Cln2 with Cdc28, or the phosphorylation on Thr-18 or Tyr-19 of Cdc28.

Initially, it was thought that only Cdc28–Cln2 complexes would be targeted by Far1, because a $CLN1^+$ $cln2$-Δ $CLN3^+$ $far1$-Δ (wild type for the CLN1 and CLN3 genes and deleted for the $CLN2$ and $FAR1$ genes) arrested in response to pheromone. Any strain containing a wild-type $CLN2$ gene and a $far1$ deletion would not arrest, however, no matter what combination of inactivating alleles of $CLN1$ or $CLN3$ were used (CHANG and HERSKOWITZ 1990). It was proposed that the activity of Cln1 and Cln3 would be reduced by other means. Deletion of $CLN2$ does not restore wild-type levels of pheromone sensitivity of $far1$-Δ strains, however, indicating that Far1 has some effect on Cln1 and/or Cln3 activity as well (CHANG and HERSKOWITZ 1990; TYERS and FUTCHER 1993). Genetic experiments have been interpreted to suggest that Far1 is active against Cln1, and not Cln3 (CHANG and HERSKOWITZ 1992; VALDIVIESO et al. 1993) or against Cln3, and not Cln1 (CHANG and HERSKOWITZ 1990). Activity assays against Cdc28–Cln1 complexes have not been reported, but immunoprecipitated Far1–Cdc28–Cln3 complexes appear to retain full protein kinase activity (TYERS and FUTCHER 1993), suggesting that Cln3 is not a target. Additional in vitro experiments are needed to clarify these issues.

McKINNEY and CROSS (1995) have presented evidence that Far1 weakly inhibits cell cycle progression in G_2 phase. Exposure of cells expressing a proteolytically stabilized allele of Far1 ($FAR1$-Δ^{1-50}) to high concentrations of pheromone (100 nM) immediately after Start causes the cells to arrest with large buds and single nuclei containing replicated DNA. This evidence has been interpreted to suggest that the G_1-phase cyclins may have a post-Start role, but could also indicate that Far1 has weak anti-Clb–Cdc28 kinase activity. PETER and HERSKOWITZ (1994) did not see any effect of Far1 on Cdc28–Clb2 and Cdc28–Clb5 complexes in their in vitro experiments, however. It is not clear whether these observations by MCKINNEY and CROSS (1995) have in vivo relevance in normal cells.

4.2 Regulation of Far1 Levels

The transcription of the *FAR1* gene is enhanced four- to fivehold induced by mating pheromone and, like other mating pheromone-induced genes, *FAR1* requires the Ste12 DNA-binding protein for its induction (CHANG and HERSKOWITZ 1990). Sequences matching the pheromone response element bound by Ste12 have been noted at positions −215, −192, −4, and +68 relative to the Far1 translation initiation site (CHANG and HERSKOWITZ 1990; KUO and GRAYHACK 1994). *FAR1* is expressed in haploid cells, but not in diploids, which do not undergo the mating pheromone response. *FAR1* is also periodically expressed during the cell cycle with mRNA levels high from early G_2 through M phase to Start and low from Start to S phase (MCKINNEY et al. 1993). Its uninduced expression correlates inversely with the expression pattern of *CLN1* and *CLN2*, the genes encoding its primary targets for inhibition. The Mcm1 transcription regulator has high-affinity binding sites at positions +189 and +211, and a third site at −262 has also been noted (KUO and GRAYHACK 1994). Mcm1 is known to be required for periodic transcriptional regulation of genes expressed late in G_2 and in M phase (LYDALL et al. 1991; ALTHOEFER et al. 1995; MAHER et al. 1995) and is also known to interact with Ste12 in the regulation of pheromone-inducible transcripts (ERREDE and AMMERER 1989; KIRKMAN-CEREIA et al. 1993).

The levels of Far1 protein also oscillate periodically with the cell cycle, but with a pattern distinct from that of its mRNA (MCKINNEY et al. 1993). Far1 is only seen in pre-Start G_1-phase cells, a pattern that is not changed when the normal Far1 promoter is replaced by the upstream activating sequence of *CYC1* which drives constitutive transcription. Additionally, cells arrested in mitosis with the microtubule inhibitor nocodazole accumulate high levels of *FAR1* mRNA and have normal bulk protein synthetic rates, but have low amounts of Far1 protein.

The discrepancy between the periodicities in Far1 protein levels and mRNA levels is best explained by cyclical Far1 proteolysis. MCKINNEY and CROSS (1995) found that Far1 is stable in G_1 phase, but must be resynthesized after S-phase entry to allow arrest. Ubiquitin-mediated proteolysis has been implicated in Far1 stability, since Far1 accumulates at the restrictive temperature in $cdc34^{ts}$ strains (MCKINNEY et al. 1993). Consistent with the putative requirement for Cdc28–Cln phosphorylation on Cdc34-dependent ubiquitination, Far1 levels are also high in cells lacking all three *CLN* genes or in $cdc28^{ts}$ strains at the restrictive temperature. The Far1 that accumulates in $cdc34^{ts}$ is hyperphosphorylated (MCKINNEY et al. 1993).

At least one determinant of Far1 stability resides in the first 30 residues of the protein, since deletions of this segment increase Far1 half-life (MCKINNEY and

CROSS 1995). This segment contains an epitope that is cell cycle regulated: an antibody raised against residues 10–24 does not recognize hyperphosphorylated Far1 soon after Start. The first 30 residues of Far1 are not just simply a degradation conferring sequence, however, since deletion of the first 40 amino acids decreases Far1 stability, and a β-galactosidase fusion with the first 50 amino acids of Far1 does not confer instability to the heterologous protein. Given these properties, this should be a valuable system for the exploration of cell cycle-associated proteolysis.

4.3 Regulation of Far1 Activity

Expression of $FAR1\text{-}\Delta^{1-50}$ (which encodes a stabilized form of Far1 lacking the first 50 amino acids) using the powerful $GAL1$ in a $far1$-deficient ($far1\text{-}\Delta$) background results in higher levels of Far1 protein than can be obtained with pheromone induction, yet cell division arrest is not observed unless the cells are also exposed to mating pheromone (CHANG and HERSKOWITZ 1992). These results suggest that Far1 needs to be activated in some way before it can inhibit Cdc28–Cln complexes. Supporting this interpretation are results implied, but not explicitly discussed, by PETER and HERSKOWITZ (1994). In their in vitro demonstration of Far1 CKI activity, they used Far1 isolated from pheromone-treated yeast cells, and not from untreated cells, and from bacteria expressing a constitutively active form of Far1 (Far1-22), and not from wild-type Far1. The implication is that unmodified Far1 did not inhibit Cdc28–Cln complexes.

Prior results suggest that phosphorylation by the MAP kinase homologue Fus3 may be needed for Far1 inhibitory activity. As stated earlier, Far1 association with Cdc28–Cln2 is greatly enhanced by Fus3 activity (PETER et al. 1993; TYERS and FUTCHER 1993). In vitro, Fus3 immunoprecipitated from mating pheromone-treated cells, but not from untreated controls or from strains expressing a kinase-dead allele of Fus3, phosphorylates Far1 (ELION et al. 1993; PETER et al. 1993). In vivo, Far1 is a phosphoprotein in untreated cells, but is hyperphosphorylated after treatment with mating pheromone. This hyperphosphorylation is not seen in $fus3$ mutants. A subset of tryptic phosphopeptides of Far1 phosphorylated in vitro by Fus3 is found to be phosphorylated in vivo in mating pheromone-treated cells (PETER et al. 1993). Given these results, it is attractive to speculate that Far1 phosphorylation by Fus3 is needed for Far1 inhibitory activity. On the other hand, Cdc28–Cln2 is also able to phosphorylate Far1 (PETER et al. 1993; TYERS and FUTCHER 1993), so Fus3-dependent hyperphosphorylation may be a result of increased Far1–Cdc28–Cln2 association, and not the direct result of a Fus3–Far1 kinase–substrate interaction. A definitive experiment showing that only Fus3-phosphorylated Far1 is an effective Cdc28–Cln2 inhibitor has not yet been reported.

A second MAP kinase homologue called Kss1 acts in concert with Fus3 in the mating pheromone response (COURCHESNE et al. 1989). Both Fus3 and Kss1 are activated by Ste7, a MAP kinase kinase homologue, and both Fus3 and Kss1 activate Ste12, a DNA binding protein that regulates the transcriptional response to mating pheromone (ELION et al. 1991b; SONG et al. 1991; GARTNER et al. 1992). Unlike Fus3, however, Kss1 does not arrest cell cycle progression when activated (ELION et al. 1991a) and has no role in the activation of Far1 (ELION et al. 1993; PETER et al. 1993). This pathway provides an interesting example of homologous genes carrying out partial, but not wholly redundant functions.

4.4 A Non-CKI Role for *FAR1*

As might be expected from the discussion above, *far1*-Δ mutants have reduced mating efficiency (4% of wild type), a defect that is far more severe when both parents are *far1*-Δ (3×10^{-7} of wild type; CHANG and HERSKOWITZ 1990). Despite being able to continue cell division, *far1*-Δ cells still respond to the presence of mating pheromone by inducing specific transcripts and displaying an altered morphology. The cells produce large lobes connected by narrow tubes and have multiple nuclei. Diploids homozygous for mating type arise at frequent intervals due to apparent nuclear fusion events in multinucleate cells. These latter phenotypes have been ascribed to the induction of mating specific transcripts in the still cycling cell, but the mating failure itself is due to more subtle defect not related to the inability to arrest in G_1 phase.

This additional function of *FAR1* in mating was suspected from the early genetic analysis of *far1* mutants (CHANG and HERSKOWITZ 1990). The most dramatic of this early evidence was the existence of *far1* alleles with transposon insertions in the 3′ two thirds of the gene, which inefficient maters but still possessed the ability to arrest the cell cycle in response to pheromone (the Mating⁻ Arrest⁺ phenotype). CHENEVERT et al. (1994) isolated additional point mutations in *FAR1* displaying the Mating⁻ Arrest⁺ phenotype (the *far1-s* alleles) in a screen for mutants defective in pheromone-induced cell polarization. The *far1-s* mutants had normal induction of mating-specific gene transcription (including that of *far1*), were still defective in their ability to mate when forced to arrest at Start through the use of a *cdc28*[ts] mutation, and had a morphological response to pheromone exposure that appeared normal. The *far1-s* mutants, however, were unable to orient their growth toward a source of pheromone (DORER et al. 1995; VALTZ et al. 1995). This is an important behavior for mating yeast cells, since oriented growth (not division) is the only means that nonmotile yeast cells have to make contact with a potential mate that is not in direct physical contact (JACKSON and HARTWELL 1990; SEGALL 1993).

Although the directed cellular growth induced by mating pheromone was not oriented towards the mating partner in the *far1-s* mutants, it was not randomly oriented either (DORER et al. 1995; VALTZ et al. 1995). Careful mapping of the growth direction in *far1-s* cells after exposure to pheromone indicated that the growth initiated from a cell surface site adjacent to the previous site of bud emergence (VALTZ et al. 1995). Normally, haploid yeast cells bud in an axial fashion, i.e., successive buds appear adjacent to the previous bud. Brief exposure to pheromone erases this axial pattern, allowing cells to reset their growth direction toward the mating partner (MADDEN and SNYDER 1992). The *far1-s* mutants continued to initiate cell growth at the cell surface location where the bud would have appeared (VALTZ et al. 1995) and did not randomize bud sites when briefly exposed to pheromone (DORER et al. 1995). These results indicate that Far1 normally functions to help the cell "forget" its normal growth site, so that growth can be redirected towards a mating partner. This function of Far1 is separable from its CKI function, in that mutants defective in growth arrest can complement orientation-defective mutants in *trans*. Four of the five sequenced *far1-s* mutations map to the carboxy terminus (residues 646, 665, 671, and 756; VALTZ et al. 1995), a region previously found to be dispensible for CKI function (CHANG and HERSKOWITZ 1990). The remaining *far1-s* mutation is in the CKI domain in a residue (205) that is part of a conserved LIM (domain common to *lin*-11, *IsL*-1 and *mec*-3) domain. LIM domains are involved in protein–protein interactions (SCHMEICHEL and BECKERLE 1994). The LIM domain in Far1 may function to bind it to the machinery controlling the initiation of a new bud.

4.5 *FAR1* and the Cell Cycle

The interaction between Far1 and the G_1-phase cyclins has been intensely studied. It is clear that, in addition to the direct affects on Cdc28–Cln kinase activity, there are also indirect effects on *CLN* transcription (VALDIVIESO et al. 1993) which can, in turn, feed back to affect Far1 expression and activity. An attempt to summarize the important relationships among these components is shown in Fig. 3, which contains a large amount of information, but at the same time loses its ability to inform. Obviously, these genes and their gene products exist in a complex web of physical and transcriptional interactions that does not yield well to linear description. It must also be kept in mind that this web is extremely dynamic and will be affected by pheromone concentrations, length of pheromone exposure (not depicted is the pheromone adaptive mechanism, which is not as well understood), cell cycle position, nutritional status of the cell, cell size, and a host of other factors. We are reaching the point at which mathematical modeling of these interactions will be necessary to interpret the consequences of future genetic manipulations (TYSON et al. 1996).

Two additional findings are worth mentioning because they are not evident in Fig. 3 or from the preceeding discussion. First, Far1 appears to have a role in cell cycle progression in normal cycling cells not exposed to pheromone. A *far1-Δ* strain has a reduced G_1 phase relative to wild-type cells (McKINNEY and CROSS 1995). Consistent with this, cells overexpressing *FAR1* have a significant G_1 delay and are 175% larger than wild-type cells. Interestingly, cells overexpressing the *FAR1-Δ*$^{1-50}$ allele also were larger than wild-type cells (by 150%) but had fewer G_1-phase cells. The amino-terminal domain seems to contain an important determinant specifying G_1-phase arrest.

The second finding is that strains lacking all three G_1-phase cyclins can still respond to mating pheromone. A *cln1-Δ cln2-Δ cln3-Δ* strain can be kept alive if the cells are also defective in the *SIC1/BYC1* gene (EPSTEIN and CROSS 1994; B.L. SCHNEIDER et al. 1996; TYERS 1996). Cell division in these cells is still responsive to mating pheromone arrest, even when *FAR1* is deleted (TYERS 1996). It seems that another layer to the circuitry depicted in Fig. 3 remains to be described.

5 Rum1

Schizosaccharomyces pombe appears to be able to control its cell cycle with considerably fewer components than are used by *S. cerevisiae*. Like *S. cerevisiae*, *S. pombe* relies on a single CDK, Cdc2, to coordinate its mitotic cell cycle events. Only four cyclins – Cdc13, Cig1, Cig2, and Puc1 – have been described. Of these, only Cdc13 is essential, and it appears that all essential cell cycle events can be carried out with only Cdc13 present (FORSBURG and NURSE 1994; FISHER and NURSE 1996). Since fission and budding yeast lifestyles are not considerably different, *S. pombe* must make efficient use of its fewer components. This is particularly evident in the way that the only known *S. pombe* CKI, Rum1, is used to prevent premature mitosis, promote S phase, and establish G1 phase (summarized in Fig. 4).

5.1 Inhibitor of Mitosis

Rum1 (GenBank X77730) is the most recently discovered fungal CKI and the only one known in *S. pombe*. With a molecular mass of 25 kDa, it is also the smallest yeast CKI. It was isolated in a screen for genes that, when overproduced, would induce extra rounds of DNA replication (MORENO and NURSE 1994). The overreplication can be extensive, with cells having more than 16 times the normal complement of DNA, thus earning $rum1^+$ its name (*r*eplication *u*ncoupled from *m*itosis). Replication is not random, but occurs in discrete rounds. Short periods of $rum1^+$ overexpression can be used to generate cells with multiple, but complete

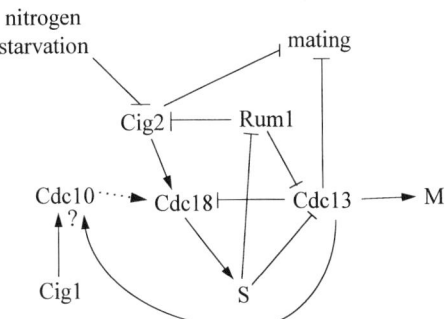

Fig. 4. Regulatory circuitry centering on Rum1. Conventions are as for Fig. 1. Cig1, Cig2, and Cdc13 are cyclins; Cdc10 is a transcription factor; Cdc18 is a putative component of DNA licensing factor; and Rum1 is a cyclin-dependent kinase (CDK) inhibitor (CKI). The cyclins are active when combined with the Cdc2 CDK, which for the sake of clarity is not shown

genomic complements. Apparently high expression of $rum1^+$ redirects cells from G_2 back into G_1 phase.

Earlier studies showed that heat-shocking $cdc2^{ts}$ alleles would induce G_2-phase cells to reenter S phase, apparently resetting the cells from G_2 to G_1 phase and implicating $cdc2^+$ in the process by which cells "remember" what phase of the cell cycle they are in (BROEK et al. 1991). The same type of extensive overreplication seen in $rum1^+$ overexpressors can also be generated by deletion or inactivation of $cdc13^+$, the gene encoding the major mitotic cyclin (HAYLES et al. 1994; FISHER and NURSE 1996). These studies suggested a role for Rum1 as an inhibitor of Cdc2–Cdc13 complexes. Consistent with this, the protein kinase activity of Cdc2–Cdc13 complexes in $rum1^+$ overexpressors was lower than in wild-type cells (MORENO and NURSE 1994), but was higher in $rum1$-Δ strains arrested in G_1 phase (CORREA-BORDES and NURSE 1995). Direct in vitro assays have now demonstrated that Rum1 in nanomolar concentrations is an effective inhibitor of Cdc2–Cdc13 activity (CORREA-BORDES and NURSE 1995; JALLEPALLI and KELLY 1996; MARTÍN-CASTE-LLANOS et al. 1996).

Additional studies support an antimitotic activity for Rum1. Rum1 accumulates in cells arrested in G_1 phase, where it is found associated with Cdc2–Cdc13 complexes. No Rum1 protein is seen in G_2-phase cells (CORREA-BORDES and NURSE 1995). More strikingly, $rum1$-Δ $cdc10^{ts}$ or $rum1$-Δ $cdc18$-$K46^{ts}$ double mutants divide at the restrictive temperature without DNA replication, producing anucleate cells or cells with improperly divided chromatin (the cut phenotype) (MORENO and NURSE 1994; JALLEPALLI and KELLY 1996). At the restrictive temperature, $cdc10$ and $cdc18$-$K46^{ts}$ mutants alone do not initiate S phase or mitosis. These experiments indicates that $rum1^+$ can act in late G_1 or early S phase to prevent premature mitosis.

5.2 Promoter of S Phase

The antimitotic activity explained why $rum1^+$ overexpressors failed to mitose following S phase, but it did not explain the failure of the controls that prevented S phase from occurring without previous mitosis. JALLEPALLI and KELLY (1996) provided a key piece of evidence suggesting that the effect of Rum1 on Cdc18 may explain the multiple S-phase initiations. Cdc18 is required for the initiation of DNA synthesis and for a checkpoint mechanism that prevents mitosis when DNA synthesis is blocked (KELLY et al. 1993). It is a labile protein that is synthesized prior to S phase and then disappears (NISHITANI and NURSE 1995; MUZI-FALCONI et al. 1996). Overexpression of $rum1^+$, however, triggered dramatic increases in the level of Cdc18 protein (JALLEPALLI and KELLY 1996). On SDS-PAGE gels, the increased Cdc18 accumulated in a slightly faster migrating band that may have been a hypophosphorylated form. In vitro Cdc2–Cdc13 complexes phosphorylated the Cdc18 amino terminus, a phosphorylation that was inhibited by Rum1. These results are consistent with a model in which phosphorylation of Cdc18 by Cdc2–Cdc13 complexes induces Cdc18 proteolysis in a manner analogous to that described for Far1, Sic1, and the Cln in *S. cerevisiae*. By inhibiting Cdc2–Cdc13, Rum1 would not only prevent mitosis, but would stabilize Cdc18 and promote S-phase initiation. This is an exciting development that clearly requires more examination.

5.3 Establishment of Pre-Start G_1 Phase

The periodicity of Rum1 has not been adequately documented, but the existing data suggests that it is primarily found in G_1-phase cells (CORREA-BORDES and NURSE 1995), where it has an important role in establishing the pre-Start phase of the cell cycle. Deletion of *rum1* results in cells with a very short G_1-phase interval, and *rum1* overexpression delays the appearance of S phase (MORENO and NURSE 1994). The cells do not arrest when starved of a nitrogen source and are unable to mate, indicating a lack of normal G_1-phase controls. In a *rum1*-Δ strain, $cdc2^{ts}$ alleles that normally arrest in both G_1 and G_2 phase leak out of G_1 and arrest in G_2 phase, indicating that the inhibitory effects of Rum1 predominate in G_1 phase (LABIB et al. 1995).

Part of the function of Rum1 in establishing G_1 phase comes as a result of its inhibition of the low levels of Cdc2–Cdc13 protein kinase activity found in G_1 phase with the consequences described above. It is also necessary for depressing the synthesis or enhancing the turnover of Cdc13 protein during the G_1-phase interval. The mechanism by which this is accomplished is unknown, but it does not involve alterations in *cdc13* transcription (CORREA-BORDES and NURSE 1995).

In addition to its interaction with Cdc2–Cdc13, Rum1 is also an inhibitor of Cdc2–Cig2 (CORREA-BORDES and NURSE 1995). Unlike Cdc13, which has peak protein levels in G_2/M phase, Cig2 is expressed primarily in G_1 phase and therefore appears to be a classical G_1-phase cyclin (MARTIN-CASTELLANOS et al. 1996). Rum1 is not as effective on Cdc2–Cig2 as it is on Cdc2–Cdc13 complexes, but the higher level of Rum1 expression apparent during G_1 phase may compensate (CORREA-BORDES and NURSE 1995).

Genetic analyses clearly indicate that *cig2* is important for G_1 phase and that *rum1* plays a part in this regulation. Deletion of *cig2* restores fertility to *rum1*-Δ, although not to wild-type levels (CORREA-BORDES and NURSE 1995). *cig2*-Δ strains are hyperfertile (CONNOLLY and BEACH 1994; OBARA-ISHIHARA and OKAYAMA 1994), so the double mutant result indicates that Cig2 has a major role in repressing mating and that *rum1* has a critical role in relieving this repression. The epistasis is probably not 100%, because deletion of *rum1* also releases Cdc2–Cdc13 activity, which could also repress mating, either directly or indirectly, by promoting mitosis.

Studies of the rereplication phenotype also support a role for *rum1* in G_1-phase events. When rereplication is caused by loss of *cdc13*, a severe delay in the onset of rereplication is observed when *cig2* is also deleted (FISHER and NURSE 1996). When rereplication is caused by overexpression of *rum1*$^+$, however, deletion of *cig2* results in no change in rereplication kinetics (MARTIN-CASTELLANOS et al. 1996). These results indicate that Cig2 is normally needed for S-phase initiation and that

Rum1 will inhibit this activity of Cig2. These experiments also demonstrated a role for $cig1^+$ in G_1 phase. Cig1 is expressed similarly to Cdc13 (G_2/M phase). When $rum1^+$ was overexpressed in a $cig1$ deletion, there was transient G_1-phase arrest and rereplication was delayed. No such effect was seen when rereplication was induced in the $cdc13$-Δ strain (FISHER and NURSE 1996), indicating that the role for Cig1 in G_1 phase is redundant with Cdc13. Since Cdc13 and Cig2, but not Cig1, are inhibited by Rum1, there may be natural events causing high Rum1 expression during which Cig1 function could be important.

6 Conclusions

An obvious function for a CKI would be to link cell cycle checkpoints to the CDK, so it is surprising that no such link has yet been found in yeast cells. This suggests that more CKI remain to be found. Sequence homology to other CKI would be the easiest route to discovery, but with the sequence of the *S. cerevisiae* genome almost complete and the *S. pombe* genome not too far behind, no obvious homologues to CKI other than Pho85 have become apparent. The Pho85 "homologues" are limited to proteins containing the ankyrin repeat domain, which, although shared with some of the mammalian CKI, is a structural unit used widely for protein–protein interactions, and not merely interactions with CDK or cyclins. It is not yet clear whether any of these have CKI activity. The other CKI – Sic1, Far1, and Rum1 – are unique at the primary sequence level. Other unique-sequence CKI can be expected to be found. Whether any of these have mammalian homologues is still an open question.

Acknowledgements. The author would like to thank Jennifer Myka and Keith Filer for thoughtful suggestions on this manuscript.

References

Althoefer H, Schleiffer A, Wassman K, Nordheim A, Ammerer G (1995) Mcm1 is required to coordinate G_2-specific transcription in Saccharomyces cerevisiae. Mol Cell Biol 15:5917–5928
Bardwell L, Cook JG, Inouye CJ, Thorner J (1994) Signal propagation and regulation in the mating pheromone response pathway of the yeast Saccharomyces cerevisiae. Dev Biol 166:363–379
Basco RD, Segal MD, Reed SI (1995) Negative regulation of the G_1 and G_2 by S-phase cyclins of Saccharomyces cerevisiae. Mol Cell Biol 15:5030–5042
Breeden L (1996) Start-specific transcription in yeast. In: Farnham PJ (ed) Current topics in microbiology and immunology, vol 208. Springer, Berlin Heidelberg New York, pp 95–127
Broek D, Bartlett R, Crawford K, Nurse P (1991) Involvement of $p34^{cdc2}$ in establishing the dependency of S phase on mitosis. Nature 349:388–393
Bun-Ya M, Nishimura M, Harashima S, Oshima Y (1991) The PHO84 gene of Saccharomyces cerevisiae encodes an inorganic phosphate transporter. Mol Cell Biol 11:3229–3238

Chang F, Herskowitz I (1990) Identification of a gene necessary for cell cycle arrest by a negative growth factor of yeast: FAR1 is an inhibitor of a G1 cyclin, CLN2. Cell 63:999–1011

Chang F, Herskowitz I (1992) Phosphorylation of FAR1 in response to α-factor: a possible requirement for cell-cycle arrest. Mol Biol Cell 3:445–450

Chenevert J, Valtz N, Herskowitz I (1994) Identification of genes required for normal pheromone-induced cell polarization in Saccharomyces cerevisiae. Genetics 136:1287–1297

Clay FJ, McEwen SJ, Bertoncello I, Wilks AF, Dunn AR (1993) Identification and cloning of a protein kinase-encoding mouse gene, Plk, related to the polo gene of Drosophila. Proc Natl Acad Sci USA 90:4882–4886

Connolly T, Beach D (1994) Interaction between the Cig1 and Cig2 B-type cyclins in the fission yeast cell cycle. Mol Cell Biol 14:768–776

Correa-Bordes J, Nurse P (1995) $p25^{rum1}$ orders S phase and mitosis by acting as an inhibitor of the $p34^{cdc2}$ mitotic kinase. Cell 83:1001–1009

Courchesne WE, Kunisawa R, Thorner J (1989) A putative protein kinase overcomes pheromone-induced arrest of cell cycling in S. cerevisiae. Cell 58:1107–1119

Deshaies RJ, Chau V, Kirschner M (1995) Ubiquitination of the G_1 cyclin Cln2p by a Cdc34p-dependent pathway. EMBO J 14:303–312

Dirick L, Böhm T, Nasmyth K (1995) Roles and regulation of Cln-Cdc28 kinases at the start of the cell cycle of Saccharomyces cerevisiae. EMBO J 14:4803–4813

Donovan JD, Toyn JH, Johnson AL, Johnston LH (1994) p40SDB25, a putative CDK inhibitor, has a role in the M/G1 transition in Saccharomyces cerevisiae. Genes Dev 8:1640–1653

Dorer R, Pryciak PM, Hartwell LH (1995) Saccharomyces cerevisiae cells execute a default pathway to select a mate in the absence of pheromone gradients. J Cell Biol 131:845–861

Elion EA, Brill JA, Fink GR (1991a) Functional redundancy in the yeast cell cycle: FUS3 and KSS1 have both overlapping and unique functions. Cold Spring Harbor Symp Quant Biol 61:41–49

Elion EA, Brill JA, Fink GR (1991b) FUS3 represses CLN1 and CLN2 and in concert with KSS1 promotes signal transduction. Proc Natl Acad Sci USA 88:9392–9396

Elion EA, Satterberg B, Kranz JE (1993) FUS3 phosphorylates multiple components of the mating signal transduction cascade: evidence for STE12 and FAR1. Mol Biol Cell 4:495–510

Epstein CB, Cross FR (1992) CLB5: a novel B cyclin from budding yeast with a role in S phase. Genes Dev 6:1695–1706

Epstein CB, Cross FR (1994) Genes that can bypass the CLN requirement for Saccharomyces cerevisiae cell cycle START. Mol Cell Biol 14:2041–2047

Errede B, Ammerer G (1989) STE12, a protein involved in cell-type-specific transcription and signal transduction in yeast is part of protein-DNA complexes. Genes Dev 3:1349–1361

Fenton B, Glover DM (1993) A conserved mitotic kinase active at late anaphase-telophase in syncytial Drosophila embryos. Nature 363:637–640

Fisher DL, Nurse P (1996) A single fission yeast mitotic cyclin B $p34^{cdc2}$ kinase promotes S-phase and mitosis in the absence of G_1 cyclins. EMBO J 15:850–860

Fisher F, Jayaraman PS, Goding CR (1991) c-myc and the yeast transcription factor PHO4 share a common CACGTG-binding motif. Oncogene 7:1099–1104

Forsburg SL, Nurse P (1994) Analysis of the Schizosaccharomyces pombe cyclin puc1: evidence for a role in cell cycle exit. J Cell Sci 107:601–613

Gartner A, Nasmyth K, Ammerer G (1992) Signal transduction in Saccharomyces cerevisiae requires tyrosine and threonine phosphorylation of FUS3 and KSS1. Genes Dev 6:1280–1292

Goebl MG, Yochem J, Jentsch S, McGrath JP, Varshavsky A, Byers B (1988) The yeast cell cycle gene CDC34 encodes a ubiquitin-conjugating enzyme. Science 241:1331–1335

Hall M, Bates S, Peters G (1995) Evidence for different modes of action of cyclin-dependent kinase inhibitors: p15 and p16 bind to kinases, p21 and p27 bind to cyclins. Oncogene 11:1581–1588

Harper JW, Elledge SJ, Keyomarsi K, Dynlacht B, Tsai L-H, Zhang P, Dobrowolski S, Bai C, Connell-Crowley L, Swindell E, Fox MP, Wei N (1995) Inhibition of cyclin-dependent kinases by p21. Mol Biol Cell 6:387–400

Hayles J, Fisher D, Woollard A, Nurse P (1994) Temporal order of S phase and mitosis in fission yeast is determined by the state of the $p34^{cdc2}$-mitotic complex. Cell 78:813–822

Hereford LM, Hartwell LH (1974) Sequential gene function in the initiation of S. cerevisiae DNA synthesis. J Mol Biol 84:445–461

Herskowitz I (1995) MAP kinase pathways in yeast: for mating and more. Cell 80:187–197

Hirst K, Fisher F, McAndrew PC, Goding CR (1994) The transcription factor, the Cdk, its cyclin and their regulator: directing the transcriptional response to a nutritional signal. EMBO J 13:5410–5420

Jackson CL, Hartwell LH (1990) Courtship in S. cerevisiae: both cell types choose mating partners by responding to the strongest pheromone signal. Cell 63:1039–1051

Jallepalli PV, Kelly TJ (1996) Rum1 and Cdc18 link inhibition of cyclin-dependent kinase to the initiation of DNA replication in Schizosaccharomyces pombe. Genes Dev 10:541–552

Johnston LH, Eberly SL, Chapman JW, Araki H, Sugino A (1990) The product of the Saccharomyces cerevisiae cell cycle gene DBF2 has homology with protein kinases and is periodically expressed in the cell cycle. Mol Cell Biol 10:1358–1366

Kaffman A, Herskowitz I, Tjian R, O'Shea EK (1994) Phosphorylation of the transcription factor PHO4 by a cyclin-CDK complex, PHO80-PHO85. Science 263:1153–1156

Kelly TJ, Martin GS, Forsburg SL, Stephen RJ, Russo A, Nurse P (1993) The fission yeast $cdc18^+$ gene product couples S phase to START and mitosis. Cell 74:371–382

Kirkman-Cerreia C, Stroke IL, Fields S (1993) Functional domains of the yeast STE12 protein, a pheromone-responsive transcriptional activator. Mol Cell Biol 13:3765–3772

Kitada K, Johnson AL, Johnston LH, Sugino A (1993) A multicopy suppressor of the Saccharomyces cerevisiae G_1 cell cycle gene dbf4 encodes a protein kinase and is identified as CDC5. Mol Cell Biol 13:4445–4457

Kühne C, Linder P (1993) A new pair of B-type cyclins from Saccharomyces cerevisiae that function early in the cell cycle. EMBO J 12:3437–3447

Kuo M, Grayhack E (1994) A library of yeast genomic MCM1 binding sites contains genes involved in cell cycle control, cell wall and membrane structure, and metabolism. Mol Cell Biol 14:348–359

Labib K, Moreno S, Nurse P (1995) Interaction of cdc2 and rum1 regulates Start and S-phase in fission yeast. J Cell Sci 108:3285–3294

Lanker S, Valdivieso DH, Wittenberg C (1996) Rapid degradation of the G_1 cyclin Cln2 induced by CDK-dependent phosphorylation. Science 271:5255

Lemire JM, Willcocks T, Halvorson HO, Bostian KA (1985) Regulation of repressible acid phosphatase gene transcription in Saccharomyces cerevisiae. Mol Cell Biol 8:2131–2141

Lew DJ, Reed SI (1993) Morphogenesis in the yeast cell cycle: regulation by Cdc28 and cyclins. J Cell Biol 120:1305–1320

Lydall D, Ammerer G, Nasmyth K (1991) A new role for MCM1 in yeast: cell cycle regulation of SWI5 transcription. Genes Dev 5:2405–2419

Madden K, Snyder M (1992) Specification of sites for polarized growth in Saccharomyces cerevisiae and the influence of external factors on site selection. Mol Cell Biol 3:1025–1035

Madden SL, Johnson DL, Bergman LW (1990) Molecular and expression analysis of the negative regulators involved in the transcriptional regulation of acid phosphatase production in Saccharomyces cerevisiae. Mol Cell Biol 10:5950–5957

Maher M, Cong F, Kindelberger D, Nasmyth K, Dalton S (1995) Cell cycle-regulated transcription of the CLB2 gene is dependent on Mcm1 and a ternary complex factor. Mol Cell Biol 6:3129–3137

Martín-Castellanos C, Labib K, Moreno S (1996) B-type cyclins regulate G_1 progression in fission yeast in opposition to the $p25^{rum1}$ cdk inhibitor. EMBO J 15:839–849

McKinney JD, Cross FR (1995) FAR1 and the G_1 phase specificity of cell cycle arrest by mating factor in Saccharomyces cerevisiae. Mol Cell Biol 15:2509–2516

McKinney JD, Chang F, Heintz N, Cross FR (1993) Negative regulation of FAR1 at the start of the yeast cell cycle. Genes Dev 7:833–843

Mendenhall MD (1993) An inhibitor of $p34^{CDC28}$ protein kinase activity from Saccharomyces cerevisiae. Science 259:216–219

Mendenhall MD, Jones CA, Reed SI (1987) Dual regulation of the yeast CDC28-p40 protein kinase complex: cell cycle, pheromone, and nutrient limitation effects. Cell 50:927–935

Mendenhall MD, Al-jumaily W, Nugroho TT (1995) The Cdc28 inhibitor $p40^{Sic1}$. In: Meijer L, Guidet S, Tung HYL (eds) Progress in cell cycle research, vol 1. Plenum, New York

Moreno S, Nurse P (1994) Regulation of progression through the G_1 phase of the cell cycle by the $rum1^+$ gene. Nature 367:236–242

Morishita T, Mitsuzawa H, Nakafuku M, Nakamura S, Hattori S, Anraku Y (1995) Requirement of Saccharomyces cerevisiae Ras for completion of mitosis. Science 270:1213–1215

Muzi-Falconi M, Brown GW, Kelly TJ (1996) $cdc18^+$ regulates initiation of DNA replication in Schizosaccharomyces pombe. Proc Natl Acad Sci USA 93:1666–1670

Nishitani H, Nurse P (1995) $p85^{cdc18}$ has a major role controlling the initiation of DNA replication in fission yeast. Cell 83:397–405

Nugroho TT, Mendenhall MD (1994) An inhibitor of yeast cyclin-dependent protein kinase plays an important role in ensuring the genomic integrity of daughter cells. Mol Cell Biol 14:3320–3328

Obara-Ishihara T, Okayama H (1994) A B-type cyclin negatively regulates conjugation via interacting with cell cycle 'start' genes in fission yeast. EMBO J 13:1863–1872

Ogawa N, Noguchi K-i, Yamashita Y, Yasuhara T, Hayashi N, Yoshida K, Oshima Y (1993) Promoter analysis of the PHO81 gene encoding a 134 kDa protein bearing ankyrin repeats in the phosphatase regulon of Saccharomyces cerevisiae. Mol Gen Genet 238:444–454

Ogawa N, Noguchi K-i, Sawai H, Yamashita Y, Yompakdee C, Oshima Y (1995) Functional domains of Pho81p, an inhibitor of Pho85p protein kinase, in the transduction pathway of P_i signals in Saccharomyces cerevisiae. Mol Cell Biol 15:997–1004

O'Neill EM, Kaffman A, Jolly ER, O'Shea EK (1996) Regulation of PHO4 nuclear localization by the PHO80-PHO85 cyclin-CDK complex. Science 271:209–212

Oshima Y (1982) Regulatory circuits for gene expression: the metabolism of galactose and phosphate. In: Strathern JN, Jones EW, Broach JR (eds) The molecular biology of the yeast Saccharomyces. Metabolism and gene expression. Cold Spring Harbor Laboratories, Cold Spring Harbor, pp 159–180

Peter M, Herskowitz I (1994) Direct inhibition of the yeast cyclin-dependent kinase Cdc28-Cln by Far1. Science 265:1228–1231

Peter M, Gartner A, Horecka J, Ammerer G, Herskowitz I (1993) FAR1 links the signal transduction pathway to the cell cycle machinery in yeast. Cell 73:747–760

Pringle JR, Hartwell LH (1981) The Saccharomyces cerevisiae cell cycle. In: Strathern JN, Jones EW, Broach JR (eds) The molecular biology of the yeast Saccharomyces. Life cycle and inheritance. Cold Spring Harbor Laboratory, Cold Spring Harbor, pp 97–142

Reed SI, Hadwiger JA, Lörincz AT (1985) Protein kinase activity associated with the product of the yeast cell division cycle gene CDC28. Proc Natl Acad Sci USA 82:4055–4059

Schmeichel KL, Beckerle MC (1994) The LIM domain is a modular protein-binding interface. Cell 79:211–219

Schneider BL, Yang Q-H, Futcher AB (1996) Linkage of replication to Start by the Cdk inhibitor Sic1. Science 272:560–562

Schneider KR, Smith RL, O'Shea EK (1994) Phosphate-regulated inactivation of the kinase PHO80-PHO85 by the CDK inhibitor PHO81. Science 266:122–126

Schwob E, Nasmyth K (1993) CLB5 and CLB6, a new pair of B cyclins involved in DNA replication in Saccharomyces cerevisiae. Genes Dev 7:1160–1175

Schwob E, Böhm T, Mendenhall MD, Nasmyth K (1994) The B-type cyclin kinase inhibitor p40^{SIC1} controls the G1 to S transition in S. cerevisiae. Cell 79:233–244

Segall JE (1993) Polarization of yeast cells in spatial gradients of α mating factor. Proc Natl Acad Sci USA 90:8332–8336

Song O, Dolan JW, Yuan YO, Fields S (1991) Pheromone-dependent phosphorylation of the yeast STE12 protein correlates with transcriptional activation. Genes Dev 5:741–750

Surana U, Amon A, Dowzer C, McGrew J, Byers B, Nasmyth K (1993) Destruction of the CDC28/CLB mitotic kinase is not required for the metaphase to anaphase transition in budding yeast. EMBO J 12:1969–1978

Toh-e A, Ueda Y, Kakimoto S, Oshima Y (1973) Isolation and characterization of acid phosphatase mutants in Saccharomyces cerevisiae. J Bacteriol 113:727–738

Toh-e A, Nakamura H, Oshima Y (1976) A gene controlling the synthesis of non-specific alkaline phosphatase in Saccharomyces cerevisiae. Biochem Biophys Acta 428:182–192

Toh-e A, Tanaka K, Uesono Y, Wickner R (1988) PHO85, a negative regulator of the PHO system, is a homolog of a protein kinase, CDC28, of S. cerevisiae. Mol Gen Genet 214:162–164

Toyn JH, Johnston LH (1994) The Dbf2 and Dbf20 protein kinases of budding yeast are activated after the metaphase to anaphase cell cycle transition. EMBO J 13:1103–1113

Toyn JH, Araki H, Sugino A, Johnston LH (1991) The cell-cycle-regulated budding yeast gene DBF2, encoding a putative protein kinase, has a homologue that is not under cell-cycle control. Gene 104:63–70

Toyn JH, Johnson AL, Johnston LH (1995) Segregation of unreplicated chromosomes in Saccharomyces cerevisiae reveals a novel G_1/M-phase checkpoint. Mol Cell Biol 15:5312–5321

Tyers M (1996) The sole essential target of Cln G_1 cyclin function in yeast is the B-type cyclin-Cdk inhibitor, p40^{SIC1}. Proc Natl Acad Sci USA 93:7772–7776

Tyers M, Futcher B (1993) Far1 and Fus3 link the mating pheromone signal transduction pathway to three G_1-phase Cdc28 kinase complexes. Mol Cell Biol 13:5659–5669

Tyson JJ, Novak B, Odell GM, Chen K, Thron CD (1996) Chemical kinetic theory: understanding cell-cycle regulation. Trends Biochem Sci 21:89–96

Ueda Y, Toh-e A, Oshima Y (1975) Isolation and characterization of recessive, constitutive mutations for repressible acid phosphatase synthesis in Saccharomyces cerevisiae. J Bacteriol 122:911–922

Uesono Y, Tanaka K, Toh-e A (1987) Negative regulators of the PHO system in Saccharomyces cerevisiae: isolation and characterization of PHO85. Nucleic Acid Res 15:10299–10309

Valdivieso MH, Sugimoto K, Jahng K, Fernandes PM, Wittenberg C (1993) FAR1 is required for posttranscriptional regulation of CLN2 gene expression in response to mating pheromone. Mol Cell Biol 13:1013–1022

Valtz N, Peter M, Herskowitz I (1995) FAR1 is required for oriented polarization of yeast cells in response to mating hormones. J Cell Biol 131:863–873

Vogel K, Horz W, Hinnen A (1989) The two positively acting regulatory proteins PHO2 and PHO4 physically interact with PHO5 upstream activation regions. Mol Cell Biol 9:2050–2057

Yaglom J, Linskens MHK, Sadis S, Rubin DM, Futcher B, Finley D (1995) p34^{cdc28}-mediated control of Cln3 cyclin degradation. Mol Cell Biol 15:731–741

Yoshida K, Ogawa N, Oshima Y (1989) Function of the PHO regulatory genes for repressible acid phosphatase synthesis in Saccharomyces cerevisiae. Mol Gen Genet 217:40–46

Inhibitors of the Cip/Kip Family

L. HENGST and S.I. REED

1. Introduction 25
2. p21$^{Cip1/Waf1/Sdi1}$ 26
3. p27^{Kip1} 30
4. p57^{Kip2} 34
5. *Xenopus* Kip Proteins 35
6. Conclusions 35
References 36

1. Introduction

Although cyclin-dependent kinase inhibitors of the Cip/Kip family were the first to be discovered in mammalian cells, and they are the best characterized to date, fundamental mysteries remain concerning their biological roles and their regulation. The family, consisting so far of three members, is characterized by a C-terminal Cdk-inhibitory domain with a shared core homology and unrelated C-terminal domains of varying size (LEE et al. 1995; MATSUOKA et al. 1995). To date, a function has been ascribed to the C-terminal domain of only one member of the family. The C terminus of the p21 has been shown to be able to bind and inactivate the function of the DNA polymerase-δ processivity factor, PCNA (J. CHEN et al. 1995, 1996a, b; GOUBIN and DUCOMMUN 1995; LUO et al. 1995; NAKANISHI et al. 1995b; WARBRICK et al. 1995; WAGA et al. 1994).

Biochemical studies have revealed that the Cdk-inhibitory domain is essential and sufficient for inhibiting cyclin-dependent kinase activities and that interactions with both the Cdk and cyclin moieties are essential for strong binding (HALL et al. 1995; HARPER et al. 1995; LUO et al. 1995; POLYAK et al. 1995b). Determination of the three-dimensional structure of the kinase-inhibitory domain of one of the Cip/Kip inhibitors, p27^{Kip1}, bound to cyclin A–Cdk2 has provided a rational basis for these findings in that the inhibitor interacts with both cyclin and Cdk (RUSSO et al. 1996). In vitro studies have been used to claim that not all Cdk–cyclin combina-

---

Department of Molecular Biology, MB-7, The Scripps Research Institute, 10550 North Torrey Pines Road, La Jolla, CA 92037, USA

tions are inhibited with equal efficiency (HARPER et al. 1995). However, these conclusions need to be considered cautiously due to the semi-quantitative nature of the experiments. In addition, studies with recombinant cyclin-dependent kinases must be extrapolated to in vivo situations with care, since other proteins could potentially modify the potency and specificity of these inhibitors.

It has been shown that Cip/Kip-type inhibitors accumulate and inhibit Cdk activities in response to a broad range of antiproliferative stimuli. Nevertheless, mice nullizygous for at least two of the inhibitors are viable without gross defects (see Chap 5, this volume; DENG et al. 1995; FERO et al. 1996; KIYOKAWA et al. 1996; NAKAYAMA et al. 1996). Furthermore, to date there is no evidence that members of the Cip/Kip family (or their lack) have a role in human malignancy, as has been shown for p16, a member of the INK4 family of CDK inhibitors (for a review, see Chap 3, this volume; HIRAMA and KOEFFLER 1996; POLLOCK et al. 1996). There is, therefore, some doubt concerning the prominence of the role these molecules play in mediating important cell cycle-regulatory functions. In this chapter, we hope to present both the established facts concerning the biological and biochemical properties of Cip/Kip inhibitors as well as some outstanding issues that remain controversial. The final judgment concerning the function or functions and importance of these inhibitors remains to be made.

2. p21$^{\text{Cip1/Waf1/Sdi1}}$

p21Cip1/Waf1/Sdi1, henceforth referred to as p21, is the founding member of the Cip/Kip family. Ironically, this protein was identified almost simultaneously by a number of investigators pursuing diverse research objectives. The p21 cDNA was cloned functionally both from a senescent cell-derived library and a p53-induced library based on ability to inhibit cell proliferation (NODA et al. 1994; EL-DEIRY et al. 1993). At the same time, p21 was identified in the context of the yeast two-hybrid screen as a CDK2-interactive protein and found to be a CDK inhibitor (HARPER et al. 1993). Several other investigators purified a 21-kDa protein initially observed in ^{35}S-labeled cyclin–CDK immune complexes, showed that it was a CDK inhibitor, and went on to again clone the p21 cDNA (GU et al. 1993; XIONG et al. 1992, 1993a, b; ZHANG et al. 1993). Finally, a p53-dependent CDK inhibitor was found to accumulate in cells subjected to ionizing radiation and shown to be p21 (DULIC et al. 1994). Subsequently, p21 has been identified in a large number of cellular contexts, and numerous cellular functions have been attributed to it. Nevertheless, the actual biological role of p21 and its mechanism of action are issues that remain to be resolved.

The most convincing evidence for p21 function is in the context of the response of cells to DNA damage. p21 was observed to accumulate and associate with Cdk complexes in a p53-dependent manner in response both to ionizing and ultraviolet (UV) radiation (ARTUSO et al. 1995; BAE et al. 1995; BRUGAROLAS et al. 1995;

DULIC et al. 1994; EL-DEIRY et al. 1994; LIU and PELLING 1995; MACLEOD et al. 1995; MCDONALD et al. 1996; MEDRANO et al. 1995; PETROCELLI et al. 1996). The fact that p53-negative cells do not arrest in G_1 phase in response to genotoxic damage and do not accumulate p21 led to the hypothesis that p21 was the effector of p53-mediated G_1-phase arrest (EL-DEIRY et al. 1993; HARPER et al. 1993). However, this view needs to be moderated in view of recent results with nullizygous mice. Fibroblasts cultured from p53-nullizygous embryos are completely lacking in the G_1-phase response to DNA damage, as expected (BATES and VOUSDEN 1996; DENG et al. 1995; KASTAN et al. 1992), whereas p21-nullizygous fibroblasts still maintain an ability, albeit impaired, to arrest in G_1 phase in response to DNA damage (BRUGAROLAS et al. 1995; DENG et al. 1995). Therefore, it appears that induction of p21 mediates a major component, but not all of the p53-dependent response. A similar behavior in response to irradiation was observed in human p21-deficient cells (WALDMAN et al. 1995). In addition, these cells lack the G_1-phase DNA-damage checkpoint when exposed to the DNA-damaging agent adriamycin, establishing p21 as the critical mediator in this p53-dependent response (WALDMAN et al. 1995). In addition, another p53-dependent response, G_1-phase arrest in response to depletion of nucleotide pools, is completely lacking in p21-nullizygous mouse embryo fibroblasts. Such fibroblasts behave similarly to p53-nullizygous fibroblasts when treated with n-phospho-n-acetyl-L-aspartate (PALA), an inhibitor of pyrimidine biosynthesis (DENG et al. 1995).

Another p53-mediated function, the maintenance of genome stability, does not appear to be dependent on p21: p21-nullizygous mouse embryo fibroblasts appear to maintain a stable karyotype in contrast to p53-nullizygous fibroblasts (DENG et al. 1995). This finding is consistent with the fact that, whereas the p53 gene is an important locus associated with human malignancy (HOLLSTEIN et al. 1991; LEVINE et al. 1991), the p21 gene apparently does not map within a tumor suppressor locus and p21-nullizygous mice do not appear to be prone to cancer (DENG et al. 1995). Thus, although p21 is a component of the response to DNA damage, this response is not critical for the maintenance of genome stability. Although a p21 polymorphism (codon 31, Ser → Arg) exists in humans (BHATIA et al. 1995; SUN et al. 1995), no mutant alleles of p21 have been recovered from human tumors (BARBOULE et al. 1995; SHIOHARA et al. 1994; VIDAL et al. 1995; WATANABE et al. 1995), except for truncation mutations in the amino terminus of the protein, which are expected to be loss of function alleles, obtained from a subset of prostate cancers (GAO et al. 1995), and a single but heterozygous mutant allele isolated from a Burkitt's lymphoma, where both the wild-type and the mutated p21 mRNA were expressed (BHATIA et al. 1995).

Because expression of p21 is highly modulated during the course of development, it has been proposed that exit from the cell cycle during terminal differentiation is mediated in some tissues by p21 (MACLEOD et al. 1995; PARKER et al. 1995). Consistent with this, the patterns of p21 expression in wild-type and p53-nullizygous embryos are identical, indicating that p21 is under alternative control during development (PARKER et al. 1995). Furthermore, it has been shown in a tissue culture model of myogenesis that p21 accumulates in a p53-independent manner as myocytes exit the cell cycle and fuse to form myotubes (GUO et al. 1995; HALEVY et al. 1995; PARKER et al. 1995; SKAPEK et al. 1995). Consistent with this expression pattern, the p21 promoter region was shown to contain elements known to respond to myogenic transcription factors, such as MyoD (GUO et al. 1995; HALEVY et al. 1995). Similar observations have been made for other differentiation models. It has been shown recently, for example, that p21 expression promoted monocyte/macrophage specific differentiation in myelomonocytic U937 cells, suggesting that inhibition of CDK is sufficient to induce cells into a specific differentiation program (LIU et al. 1996). However, the fact that the p21-nullizygous mice develop normally excludes any critical, non-redundant role for p21 during the course of development and suggests that experiments with tumor cell-derived models need to be interpreted with caution. It is, of course, possible that such mice adjust to the lack of one regulator early in development by utilizing another in a mode not occurring in normal animals. Alternatively, redundant mechanisms are employed to mediate exit from the cell cycle in the context of terminal differentiation in intact organisms.

p21 has also been shown to accumulate as cell cultures age and approach senescence (NODA et al. 1994). However, the biological relevance of this is not clear, since there is no evidence that cells from p21-nullizygous mice become immortal more easily than controls. This is in contrast to fibroblasts from mice nullizygous for the INK4 family inhibitor, p16, which appear to be immortal at the population level when isolated from embryos (SERRANO et al. 1996). The correspondence between loss of p16 expression and susceptibility to malignancy and lack of such a correlation with loss of p21 is consistent with these observations and with uncertainty concerning a role for p21 in cellular senescence.

Although the role of p21 in the response to DNA damage is clear, the biology and biochemistry of p21 function remains unresolved. Numerous in vitro mutagenesis experiments have revealed that only the amino-terminal half of the p21 molecule is required for Cdk inhibition (J. CHEN et al. 1995, 1996a, c; FOTEDAR et al. 1996; GOUBIN and DUCOMMUN 1995; HARPER et al. 1995; LUO et al. 1995; NAKANISHI et al. 1995a). This region contains all of the conserved motifs found in other members of the Cip/Kip family. In vitro studies have revealed that the amino-terminal region of conserved sequence is responsible for cyclin binding, whereas the more carboxy-terminal conserved domain is involved in Cdk interaction (CHEN et al. 1996c; FOTEDAR et al. 1996). Thus production of a high-affinity complex requires interaction with both the Cdk and cyclin moieties of the target cyclin-dependent kinase (HALL et al. 1995; HARPER et al. 1995). This conclusion has been confirmed by the determination of the three-dimensional structure of the related inhibitor, p27, by X-ray diffraction crystallography (RUSSO et al. 1996). One mystery that remains to be solved is how p21 (and other Cip/Kip family members) maintains a broad target spectrum. In particular, p21 is an efficient inhibitor of cyclin D–Cdk4 complexes as well as of cyclin A–Cdk2, cyclin E–Cdk2, and possibly cyclinB–Cdk1 (A.B. Niculescu and S.I. Reed, unpublished). This is remarkable because there is only limited primary structure homology between these various cyclins and particularly the Cdk. Cdk2 and Cdk4 are quite divergent. Therefore, it is mysterious that a single small domain can form high-affinity complexes with a spectrum of nonconserved targets. One possible explanation derives from an unusual structural feature of the p21 Cdk-inhibitory domain, as revealed from spectroscopic studies (KIWACKI et al. 1996). In the absence of a Cdk or cyclin, this domain shows no significant fixed secondary structure. However, a defined structure is assumed when it is bound to Cdk2, suggesting that the fold conforms to the shape of the target molecule.

Another aspect of p21 biology that remains to be resolved is that of the role of interaction with proliferating cell nuclear antigen (PCNA). The carboxy terminus of p21 contains a high-affinity binding site for PCNA, which is a subunit of DNA polymerase-δ, essential for both replicative and repair DNA synthesis (J. CHEN et al. 1995; CHEN et al. 1996a, b; GOUBIN and DUCOMMUN 1995; LUO et al. 1995; NAKANISHI et al. 1995b; WAGA et al. 1994). The molar ratio of p21 to PCNA in human fibroblasts is close to 1:1 (LI et al. 1996), and binding of a p21 peptide to each PCNA molecule in the homotrimeric ring of PCNA occurs without interfering with the central hole involved in DNA interaction, as determined by X-ray crystallographic analysis (GULBIS et al. 1996). Furthermore, it has been shown in in vitro model systems that, by binding PCNA, p21 can inhibit replicative DNA synthesis, whereas the inhibition of repair synthesis by p21 needs further clarification (LI et al. 1994; PAN et al. 1995; PODUST et al. 1995; SHIVJI et al. 1994). Immunofluorescence studies of cells responding to UV irradiation have shown that excision repair is possible under high nuclear p21 concentrations, whereas DNA replicaction is inhibited (LI et al. 1996). However, it is not yet clear what role the anti-PCNA activity of p21 plays in the cell cycle response to DNA damage. Overexpression of the C-terminal domain of p21 in mammalian cells was sufficient to inhibit DNA replication (J. CHEN et al. 1995; GOUBIN and DUCOMMUN 1995; LUO et al. 1995; WARBRICK et al. 1995), while transfection of the PCNA-binding domain into fibroblasts did not appear to confer cell cycle arrest (NAKANISHI et al. 1995). However, it is possible that mobilization of the anti-PCNA activity of p21 in vivo requires concomitant activation of other responses to DNA damage not initiated when p21 is simply ectopically expressed.

A final outstanding issue is that of the stoichiometry of p21 function. It has been reported, based on in vitro experiments, that a ratio of two p21 molecules per Cdk is required for inhibition, whereas a single molecule had no inhibitory effect (HARPER et al. 1995; ZHANG et al. 1994a, b). Furthermore, in extracts of fibroblasts, the vast majority of the cyclin–Cdk2 complexes could be immunodepleted with anti-p21 antibodies (HARPER et al. 1995). These studies led to the proposal and broad acceptance of a model in which Cdk complexes are primed in a noninhibitory state through the cell cycle with single molecules of p21, and increase of p21 stoichiometry through induction of p21 expression promotes conversion to the inhibitory state. The possible advantage of having Cdk complexes primed with p21 has, however, never been convincingly articulated. It has been proposed that p21 in the noninhibitory state might have other functions, such as serving as an assembly or targeting factor. However, several observations put the original hypothesis in doubt. First, p21 in cycling cells is not detected in the nucleus during S phase and most of G_2 phase (V. Dulic and S.I. Reed, unpublished). This precludes constitutive association with Cdk complexes known to localize to the nucleus during S phase and/or G_2 phase, such as cyclin E–Cdk2 and cyclin A–Cdk2. Furthermore, in vitro inhibition studies performed more recently by other investigators suggest that binding of a single molecule of p21 to active Cdk complexes is inhibitory, in conflict with the initial reports (L. Hengst and S.I. Reed, unpublished). Certainly, it is clear by extrapolation from the structural determination of p27 bound to cyclin A–Cdk2 that one Cip/Kip inhibitor molecule should be sufficient for inhibition of a kinase molecule. These issues therefore need to be resolved before a comprehensive understanding of p21 function can be attained.

3. p27^{Kip1}

Cdk-inhibitory activity of p27^{Kip1} (*k*inase-*i*nhibitory *p*rotein), subsequently referred to as p27, was originally detected in mink lung epithelial cells undergoing transforming growth factor (TGF)-β-mediated or cell–cell contact-induced growth arrest (POLYAK et al. 1994a; SLINGERLAND et al. 1994) and in HeLa cells during G_1-phase progression and drug-mediated growth arrest by lovastatin (HENGST et al. 1994). A cDNA encoding p27 was isolated using two different approaches: (1) by a yeast interaction screen using cyclin D1–Cdk4 as bait (TOYOSHIMA and HUNTER 1994), and (2) using protein sequence information obtained by protein purification from contact-inhibited mink lung epithelial cells or from lovastatin-arrested HeLa cells (HENGST and REED 1996; POLYAK et al. 1994b). Cloning of the p27 genomic sequences revealed that the gene is interrupted by two introns, one of them being located at position 474 of the coding region of the cDNA. The position of this intron is conserved in the human p21 and p27 genes (FERRANDO et al. 1995; KAWAMATA et al. 1995; PIETENPOL et al. 1995), and the second is located in the 3′-noncoding region. A polymorphism in the human sequence was discovered at codon 109, resulting in a nonconservative amino acid substitution that changes a valine to glycine at a population frequency of 10%–20% (KAWAMATA et al. 1995; PIETENPOL et al. 1995).

As negative regulators of Cdk kinase activity, Cdk inhibitors might be expected to correspond to tumor suppressors. However, attempts to link p27 to cancer have been unsuccessful. The p27 gene was mapped to the short arm of chromosome 12, at the 12p12–12p13 boundary, a region that has been reported to harbor deletions and rearrangements in various tumors (BULLRICH et al. 1995; PIETENPOL et al. 1995; PONCE et al. 1995; TAKEUCHI et al. 1996). However, in all human cancers analyzed to date, the p27 gene seems not to be the tumor suppressor gene implicated at this locus, since no alterations were found in the p27 coding region (BULLRICH et al. 1995; FERRANDO et al. 1995; KAWAMATA et al. 1995, 1996; PIETENPOL et al. 1995; PONCE et al. 1995; TAKEUCHI et al. 1996). Moreover, p27-deficient mice are not predisposed to a general increase in tumor frequency. However, similar to mice with Rb mutations, these mice exhibit a high frequency of pituitary tumors (FERO et al. 1996; KIYOKAWA et al. 1996; NAKAYAMA et al. 1996). Recently, it has been reported that p27 can act as a tumor suppressor when overexpressed in a human astrocytoma cell line, preventing DNA rereplication in asynchronously growing or nocodazole-arrested cells and tumor formation in nude mice (CHEN et al. 1996d). Since the adenovirus-based expression system used in this study led to an ectopic overproduction of p27, these phenotypes are likely to reflect a gain of function and do not prove that p27 normally acts as a tumor suppressor.

Like p21, p27 is heat stable and remains active even after prolonged incubation at 100 °C. This property has been valuable for the purification of this inhibitor (HENGST and REED, 1996; POLYAK et al. 1994b).

In vitro, p27 interacts with and efficiently inhibits a broad range of Cdk–cyclin complexes, including Cdk1, Cdk2, Cdk4, and Cdk6 (HARPER et al. 1995; HENGST et al. 1994; POLYAK et al. 1994a, b; TOYOSHIMA and HUNTER 1994). However, while p27 binds to Cdk5–cyclin D_2 complexes, the same Cdk subunit in complex with the brain-specific activator p35 does not associate with p27 in vivo or in vitro (LEE et al. 1996). Similarly, the Cdk-activating kinase Cdk7–cyclin H is not bound by p27. However, association of p27 with the Cdk substrates of Cdk7–cyclin H leads to inhibition of their activating phosphorylation (APRELIKOVA et al. 1995; KATO et al. 1994).

A high resolution picture of the interaction of p27 with cyclin–Cdk complexes recently became available through the X-ray crystallographic analysis of a ternary Cdk–cyclin–p27 complex. A single 69-amino acid fragment of p27 including the inhibitory domain contacts both subunits of the phosphorylated cyclin A–Cdk2 complex. This dual interaction is likely to be initiated with the cyclin, since biochemical studies indicate that inhibitor–cyclin complexes are formed much more readily than inhibitor–Cdk complexes (HALL et al. 1995). The amino terminus of p27 binds to the cyclin subunit at a groove formed by the conserved cyclin box L-helices. The lack of interaction of p27 with both Cdk5–p35 and Cdk7–cyclin H complexes may thus be due to the absence of the p27-interacting amino acids in p35 and cyclin H (RUSSO et al. 1996). Whereas p27 association leaves the cyclin structure unchanged, the binding of the inhibitor to the amino-terminal lobe of Cdk2 leads to a rearrangement of this domain of the kinase subunit. The C-terminal portion of the inhibitory domain inserts into the catalytic cleft of Cdk2, bringing a tyrosine of the inhibitor to the purine-binding pocket of the kinase, thereby occupying the ATP-binding site and accounting for inhibition of kinase activity (RUSSO et al. 1996). The structure thus supports an interpretation that one p27 molecule per Cdk–cyclin complex is sufficient to inhibit kinase activity.

However, this interpretation has been a matter of controversy, as it has been reported that more than one molecule of the p27 homologue p21 is needed to inactivate Cdk complexes (HARPER et al. 1995; ZHANG et al. 1994a, b). As the primary structure of the Cdk- and cyclin-binding region (RUSSO et al. 1996) and most biochemical properties concerning Cdk–cyclin-binding and association of p21 and p27 are very similar (HALL et al. 1995; HARPER et al. 1995; LEE et al. 1996), it would be surprising if their mode of inhibition were significantly different. Biochemical analysis of the inhibition of cyclinA–Cdk2 kinase activity by p27 strongly supports the interpretation that only one p27 molecule is required for inhibition (L. Hengst and S.I. Reed, unpublished data).

Whereas the N terminus of p27 is involved in and sufficient for kinase inactivation (TOYOSHIMA and HUNTER 1994), much less is known about the C-terminal domain of the protein. This region might contain an additional cyclin-interaction site, since it has been reported to bind cyclin D alone (TOYOSHIMA and HUNTER 1994). In addition, p27 is targeted by the adenovirus E1A oncoprotein via a strong interaction with the carboxy-terminal region of p27 as well as a weaker interaction with its amino-terminal half (MAL et al. 1996). The binding of E1A to p27 blocks the inhibitory activity of the protein. This interaction is believed to contribute to the ability of E1A to overcome the growth-inhibitory effect of TGF-β in mink lung epithelial cells, since these cells regain cyclin E kinase activity in vivo and since E1A can activate p27-inhibited cyclin E kinase in vitro (MAL et al. 1996).

p27 protein levels have been shown to increase under conditions in which cells are growth arrested in response to antimitogenic or differentiation signals (outlined below). However, as these arrests result in cell cycle synchronization, it remains to be determined whether p27 induction observed under these circumstances contributes to the arrest or is a consequence of it. Induction of p27 protein levels has been observed in a number of tissue culture differentiation systems (HENGST and REED 1996; KRANENBURG et al. 1995; LIU et al. 1996; WANG et al. 1996). Evidence for an active role of p27 in promoting differentiation comes from the observation that ectopically expressed p27 induces cyclin D_3 expression and characteristic morphological changes observed in neuronal differentiation in mouse neuroblastoma cells (KRANENBURG et al. 1995). Moreover, myelomonocytic U937 cells induce monocyte/macrophage-specific markers when p27 (or p21) is expressed ectopically (LIU et al. 1996).

Based on tissue culture models, p27 has also been implicated as a mediator of various antimitogenic signals, including TGF-β (POLYAK et al. 1994a; REYNISDOTTIR et al. 1995; SLINGERLAND et al. 1994), rapamycin (NOURSE et al. 1994), cyclic adenosine monophosphate (cAMP; KATO et al. 1994), contact inhibition (HENGST et al. 1994; HENGST and REED 1996; POLYAK et al. 1994a; SLINGERLAND et al. 1994), growth factor deprivation (AGRAWAL et al. 1996; COATS et al. 1996; POON et al. 1995; SCHULZE et al. 1996; WINSTON et al. 1996), anti-IgM (EZHEVSKY et al. 1996), or loss of cell anchorage (FANG et al. 1996; SCHULZE et al. 1996; ZHU et al. 1996). Association of p27 with the majority of cyclin A–Cdk2 and cyclin E–Cdk2 complexes could be demonstrated in cells arrested with anti-IgM (EZHEVSKY et al. 1996) or fibroblasts deprived of mitogens (COATS et al. 1996), suggesting that the kinase inactivation observed under these conditions is a consequence of inhibition by p27. If p27 were the only factor responsible for kinase inactivation under these circumstances, inhibition of p27 function should prevent the cell cycle arrest. In fact, antisense inhibition of p27 expression in murine fibroblast cell lines prevented cell cycle arrest in response to mitogen depletion (COATS et al. 1996).

Paradoxically, mouse embryonic fibroblasts obtained from p27-nullizygous animals were able to undergo growth arrest upon serum starvation or contact inhibition (NAKAYAMA et al. 1996). In addition, T lymphocytes obtained from p27-

nullizygous animals were responsive to TGF-β and rapamycin (NAKAYAMA et al. 1996), suggesting that p27 is not essential for growth arrest in p27$^{-/-}$ animals. It remains to be determined whether and, if so, how Cdk kinase activity in cells from p27-nullizygous animals is regulated under the various arrest conditions discussed above. It is possible, for example, that other Cdk inhibitors are induced to compensate for the lack of p27 inhibition. Alternatively, other mechanisms of Cdk inactivation, for example increased inhibitory Cdk phosphorylation, might be involved in kinase inhibition in p27 nullizygous cells.

During the cell cycle, the inhibitory activity of p27 fluctuates, reaching maximal levels in G_1 phase. This activity correlates well with the levels of p27 protein, indicating that p27 activity is regulated primarily by protein abundance and not by protein modification. Maximal levels of p27 were reached during early G_1 phase and declined as cells entered S phase. This pattern of p27 protein abundance was observed in HeLa cells, HL60 human promyelocytic leukemia cells, and normal human diploid fibroblasts (HS68), suggesting that G_1/S phase-specific periodic accumulation of p27 protein is characteristic of a broad range of cell lines (HENGST and REED 1996). In contrast, it has been reported that Swiss 3T3 cells and MANCA human Burkitt's lymphoma cells show little variation of p27 protein levels during the cell cycle (POON et al. 1995; SOOS et al. 1996; TOYOSHIMA and HUNTER 1994). It remains to be seen whether these cells have become defective in regulatory mechanisms that control p27 levels.

Two functions for p27 during G_1-phase progression have been proposed. Early in G_1 phase, p27 might act as a physiological buffer to prevent premature activation of cyclin–Cdk complexes, until the kinase protein level exceeds that of p27 or until p27 levels decrease (HENGST et al. 1994; HENGST and REED 1996; POLYAK et al. 1994a). This idea is supported by the observation that the kinase activity associated with ectopically overexpressed cyclins A or E is strongly inhibited early in G_1 phase and increases as cells approach S phase. The level of cyclin A–Cdk inhibition is paralleled by levels of p27 associated with the kinase complex (RESNITZKY et al. 1995; T. Herzinger and S.I. Reed, unpublished observation). In addition, fibroblasts depleted for p27 protein have a shortened G_1 phase (COATS et al. 1996), supporting the hypothesis that p27 serves to prevent premature kinase activity that would otherwise accelerate G_1-phase progression.

In contrast to p21, p27 is generally not controlled by transcriptional mechanisms (but see below). p27 mRNA levels remain unchanged under diverse conditions in which protein induction has been observed (AGRAWAL et al. 1996; HALEVY et al. 1995; HENGST and REED 1996; POLYAK et al. 1994a). p27 abundance is regulated translationally under a variety of conditions, including upregulation of p27 levels by lovastatin or by density-mediated growth arrest or repression of p27 expression by platelet-derived growth factor (PDGF) (AGRAWAL et al. 1996; HENGST and REED 1996). It remains to be determined whether regulation of p27 translation efficiency is achieved at the level of nuclear cytoplasmic export of p27 mRNA or at the level of translational initiation.

In addition to translational control, regulation at the level of degradation by the ubiquitin/proteasome pathway plays a key role in establishing p27 levels.

Compared with proliferating cells, the level of p27 ubiquitination is decreased in quiescent cells (PAGANO et al. 1995), accounting for an observed increase in half-life (HENGST and REED 1996; PAGANO et al. 1995). Similarly, p27 in nonadherent NIH3T3 cells has an increased half-life compared to the protein of adherent cells, and extracts of these cells degrade p27 at a higher rate (SCHULZE et al. 1996). It has been demonstrated in vitro that the ubiquitinating enzymes Ubc3 and Ubc2 are capable of p27 ubiquitination (PAGANO et al. 1995).

While the level of p27 mRNA remains unchanged under various conditions in most cell types investigated, a strong increase in p27 mRNA has been described during vitamin D_3-mediated differentiation of U937 promyelocytic leukemia cells (LIU et al. 1996). Whereas the induction of p27 mRNA levels reached maximal levels within 4–8 h after induction, an increase in protein was only observed after roughly 40 h, at a time when p27 mRNA levels had already declined, consistent with a critical role for post-transcriptional regulation (LIU et al. 1996). In addition, in another promyelocytic leukemia cell line (HL60), vitamin D_3 promotes elevated p27 protein levels without changes in p27 mRNA levels (HENGST and REED 1996).

4. $p57^{Kip2}$

A protein closely related structurally to $p21^{Cip1/Waf1/Sdi1}$ and $p27^{Kip1}$, named $p57^{Kip2}$, was isolated by two different approaches: (1) as the product of a mouse cDNA detected using a DNA hybridization approach with p21 cDNA as probe (LEE et al. 1995) and (2) as a mouse protein interacting with cyclin D_1 in the yeast two-hybrid screen (MATSUOKA et al. 1995). Three different mouse p57 cDNA were isolated and represent likely products of alternative splicing. Two of these splice variants lead to a truncated protein that lacks 13 amino-terminal amino acids (LEE et al. 1995). Using the mouse p57 cDNA as hybridization probe, two human p57 cDNA have been isolated which also appear to be splice variants. The shorter cDNA encodes a truncated protein that consists primarily of the inhibitory domain, whereas the other cDNA encodes a protein similar in length to the mouse protein (MATSUOKA et al. 1995).

The amino-terminal region of p57, including the Cdk-inhibitory domain, has the greatest primary structure conservation relative to p21 and p27. No sequence homology with p21 or p27 is found in a region between amino acids 140 and 310, whereas the carboxy-terminal 40 amino acids shown to contain a nuclear localization signal share some sequence homology (LEE et al. 1995). A more extended homology region with only p27 is found at the very carboxy terminus of the protein in a domain containing a Cdk phosphorylation consensus site. Comparing the longer human protein and the mouse p57 protein, the amino- and carboxy-terminal sequences are highly conserved, whereas the intervening sequence shows a low degree of sequence identity. p57 has been shown to bind Cdk2, Cdk3, and Cdk4 in a cyclin dependent manner and is able to inhibit the activities of these kinases.

Overproduction of p57 leads to cell cycle arrest in G_1 phase in SAOS-2 and R-1B/ L17 cells (LEE et al. 1995; MATSUOKA et al. 1995). The mRNA is expressed in a tissue-specific manner, with highest levels in the placenta (LEE et al. 1995; MATSUOKA et al. 1995). However, the biological function of p57 remains to be elucidated.

The p57 gene is interrupted by one intron in codon 275 and by another intervening sequence in the 3'-untranslated region and is localized to chromosome 11p15.5 (MATSUOKA et al. 1995, 1996; REID et al. 1996). Even though this region of the genome is linked to Wilms tumor (WT2) and Beckwith-Weidemann syndrome, no indications of an involvement of p57 in human tumorgenesis have been obtained (ORLOW et al. 1996; REID et al. 1996). However, both diseases show evidence of the involvement of imprinting, with a loss of heterozygosity or translocations of only the maternal allele. Interestingly, p57 from humans or from mouse is subject to imprinting, with preferential expression of the maternal allele (HATADA and MUKAI, 1995; KONDO et al. 1996; MATSUOKA et al. 1996).

5. *Xenopus* Kip Proteins

Two inhibitors in the Kip/Cip family, $p27^{Xic1}$ and $p28^{Kix1}$, have been identified in *Xenopus*, using sequence homology to p21 and p27 and polymerase chain reaction (PCR) approaches (SHOU and DUNPHY 1996; SU et al. 1995). The frog proteins share over 90% sequence identity and interact and inhibit Cdk–cyclin complexes. In addition, both proteins bind PCNA, but they do so with a substantially lower affinity than human p21 (SHOU and DUNPHY 1996). The proteins seem to share roughly equal degrees of sequence conservation with the mammalian p21 and p27 proteins, having a C terminus characteristic of p27 and an imperfectly conserved region homologous to the PCNA-binding domain of p21. In frog oocytes, the protein concentration of p28 is low (approximately 2 n*M*), but increases about 100-fold during development between stages 12 and 13. The high level of p28 persists later in development and in somatic cells (SHOU and DUNPHY 1996).

6. Conclusions

Despite the intense experimental scrutiny received by members of the Cip/Kip family Cdk inhibitors over the past several years, these proteins remain enigmatic and present the biological research community with several important challenges. The most critical of these to determe the role of the various members of this family in the control of proliferation and tumorigenesis. Although, based both on inferences from in vitro studies and on tissue culture-based experiments, p21 and p27 might be expected to be tumor suppressors, analysis of human malignancies and nullizygous mouse models do not appear to bear this out. For p57, which maps near a locus associated with human malignancy and for which a nullizygous mouse remains to be described, the issue has not yet been resolved. Similarly, the expectation of critical roles for p21 and p27 in control of proliferation and development based on tissue culture models has had to be tempered in the face of experiments with nullizygous mice. Although both p21- and p27-nullizygous mice exhibit abnormalities, they develop, for the most part, normally. More critically, whereas cells from p21-nullizygous mice show a partial defect in their ability to arrest in G_1 phase in response to DNA damage, they do not exhibit genetic instability, as do p53-nullizygous mice. Similarly, although p27-nullizygous mice are large and show specific tissue hyperplasias, cells from these animals appear to respond appropriately to antimitogenic signals. These results suggest functional redundancy, involving possibly different members of the Cip/Kip family, or entirely distinct parallel regulatory systems. Clearly, there is much more to be learned about the functions of these proteins.

The second major challenge concerns the issue of targets and stoichiometry. Although studies with recombinant proteins have suggested a hierarchy of sensitivity to Cip/Kip inhibitors, the validity of these conclusions and their relevance to in vivo function is not yet clear. The issue has been further unsettled by determination of the co-crystal structure of p27 bound to Cdk2–cyclin A. In particular, it is not clear why Cdk1–cyclin B complexes should be less effectively inhibited than Cdk2–cyclin A complexes, as inferred from in vitro inhibition studies, since the regions of the respective proteins involved in the interactions are highly conserved. The co-crystal structure is also relevant in the matter of stoichiometry, which remains controversial. Whereas it has been proposed that p21, in particular, requires a stoichiometry of two inhibitor molecules to one Cdk complex for inhibition, the structure of p27 bound to Cdk2–cyclin A suggests that one molecule should be sufficient. Although a structure of p21 has not yet been reported, the high degree of conservation in the inhibitory domains of these two molecules, particularly in motifs important for inhibition, indicates that there will not be a significant difference between them. Nevertheless, the idea of 2:1 stoichiometry has gained virtually universal acceptance in the cell cycle field. Clearly, a rigorous reexamination of this issue is in order.

References

Agrawal D, Hauser P, McPherson F, Dong F, Garcia A, Pledger WJ (1996) Repression of p27^{kip1} synthesis by platelet-derived growth factor in BALB/c 3T3 cells. Mol Cell Biol 16:4327–4336

Aprelikova O, Xiong Y, Liu ET (1995) Both p16 and p21 families of cyclin-dependent kinase (CDK) inhibitors block the phosphorylation of cyclin-dependent kinases by the CDK-activating kinase. J Biol Chem 270:18195–18197

Artuso M, Esteve A, Bresil H, Vuillaume M, Hall J (1995) The role of the Ataxia telangiectasia gene in the p53, WAF1/CIP1(p21)- and GADD45-mediated response to DNA damage produced by ionising radiation. Oncogene 11:1427–1435

Bae I, Fan S, Bhatia K, Kohn KW, Fornace AJ, O'Connor PM (1995) Relationships between G1 arrest and stability of the p53 and p21Cip1/Waf1 proteins following gamma-irradiation of human lymphoma cells. Cancer Res 55:2387–2393

Barboule N, Mazars P, Baldin V, Vidal S, Jozan S, Martel P, Valette A (1995) Expression of p21WAF1/CIP1 is heterogeneous and unrelated to proliferation index in human ovarian carcinoma. Int J Cancer 63:611–615

Bates S, Vousden KH (1996) p53 in signalling checkpoint arrest or apoptosis. Curr Opin Genet Dev 6:12–18

Bhatia K, Fan S, Spangler G, Weintraub M, O'Connor PM, Judde JG, Magrath I (1995) A mutant p21 cyclin-dependent kinase inhibitor isolated from a Burkitt's lymphoma. Cancer Res 55:1431–1435

Brugarolas J, Chandrasekaran C, Gordon JI, Beach D, Jacks T, Hannon GJ (1995) Radiation-induced cell cycle arrest compromised by p21 deficiency. Nature 377:552–557

Bullrich F, MacLachlan TK, Sang N, Druck T, Veronese ML, Allen SL, Chiorazzi N, Koff A, Heubner K, Croce CM, et al. (1995) Chromosomal mapping of members of the cdc2 family of protein kinases, cdk3, cdk6, PISSLRE, and PITALRE, and a cdk inhibitor, p27Kip1, to regions involved in human cancer. Cancer Res 55:1199–1205

Chen IT, Akamatsu M, Smith ML, Lung FD, Duba D, Roller PP, Fornace AJ, O'Connor PM (1996a) Characterization of p21Cip1/Waf1 peptide domains required for cyclin E/Cdk2 and PCNA interaction. Oncogene 12:595–607

Chen J, Jackson PK, Kirschner MW, Dutta A (1995) Separate domains of p21 involved in the inhibition of Cdk kinase and PCNA. Nature 374:386–388

Chen J, Peters R, Saha P, Lee P, Theodoras A, Pagano M, Wagner G, Dutta A (1996b) A 39 amino acid fragment of the cell cycle regulator p21 is sufficient to bind PCNA and partially inhibit DNA replication in vivo. Nucleic Acids Res 24:1727–1733

Chen J, Saha P, Kornbluth S, Dynlacht B, Dutta A (1996c) Cyclin-binding motifs are essential for the function of p21^{CIP1}. Mol Cell Biol 19:4673–4682

Chen J, Willingham T, Shuford M, Nisen PD (1996d) Tumor suppression and inhibition of aneuploid cell accumulation in human brain tumor cells by ectopic overexpression of the cyclin-dependent kinase inhibitor p27KIP1. J Clin Invest 97:1983–1988

Coats S, Flanagan WM, Nourse J, Roberts JM (1996) Requirement of p27^{Kip1} for restriction point control of the fibroblast cell cycle. Science 272:877–880

Deng C, Zhang P, Harper JW, Elledge SJ, Leder P (1995) Mice lacking p21CIP1/WAF1 undergo normal development, but are defective in G1 checkpoint control. Cell 82:675–684

Dulic V, Kaufmann WK, Wilson SJ, Tlsty TD, Lees E, Harper JW, Elledge SJ, Reed SI (1994) p53-dependent inhibition of cyclin-dependent kinase activities in human fibroblasts during radiation-induced G1 arrest. Cell 76:1013–1023

El-Deiry W, Tokino T, Velculescu VE, Levy DB, Parsons R, Trent JM, Lin D, Mercer WE, Kinzler KW, Vogelstein B (1993) WAF1, a potential mediator of p53 tumor suppression. Cell 75:817–825

El-Deiry W, Harper JW, O'Connor PM, Velculescu VE, Canman CE, Jackman J, Pietenpol JA, Burrell M, Hill DE, Wang Y, et al. (1994) WAF1/CIP1 is induced in p53-mediated G1 arrest and apoptosis. Cancer Res 54:1169–1174

Ezhevsky SA, Toyoshima H, Hunter T, Scott DW (1996) Role of cyclin A and p27 in anti-IgM-induced G1 growth arrest of murine B-cell lymphomas. Mol Biol Cell 7:553–564

Fang F, Orend G, Watanabe N, Hunter T, Ruoslahti E (1996) Dependence of cyclin E-CDK2 kinase activity on cell anchorage. Science 271:499–502

Fero ML, Rivkin M, Tasch M, Porter P, Carow CE, Firpo E, Polyak K, Tsai LH, Broudy V, Perlmutter RM, Kaushansky K, Roberts JM (1996) A syndrome of multiorgan hyperplasia with features of gigantism, tumorigenesis, and female sterility in p27(Kip1)-deficient mice. Cell 85:733–744

Ferrando AA, Balbin M, Pendas AM, Vizoso F, Velasco G, Lopez-Otin C (1995) Mutational analysis of the human cyclin-dependent kinase inhibitor p27^{Kip1} in primary breast carcinomas. Hum Genet 967:91–94

Fotedar R, Fitzgerald P, Rousselle T, Cannella D, Doree M, Messier H, Fotedar A (1996) p21 contains independent binding sites for cyclin and cdk2:both sites are required to inhibit cdk2 kinase activity. Oncogene 12:2155-2164

Gao X, Chen YQ, Wu N, Grignon DJ, Sakr W, Porter AT, Honn KV (1995) Somatic mutations of the WAF1/CIP1 gene in primary prostate cancer. Oncogene 11:1395–1398

Goubin F, Ducommun B (1995) Identification of binding domains on the p21Cip1 cyclin-dependent kinase inhibitor. Oncogene 10:2281–2287

Gu Y, Turck CW, Morgan DO (1993) Inhibition of CDK2 activity in vivo by an associated 20K regulatory subunit. Nature 366:707–710

Gulbis JM, Kelman Z, Hurwitz J, O'Donnel M, Kuriyan, J (1996) Structure of the C-terminal region of p21$^{Waf1/Cip1}$ complexed with human PCNA. Cell 87:297–306

Guo K, Wang J, Andres V, Smith RC, Walsh K (1995) MyoD-induced expression of p21 inhibits cyclin-dependent kinase activity upon myocyte terminal differentiation. Mol Cell Biol 15:3823–3829

Halevy O, Novitch BG, Spicer DB, Skapek SX, Rhee J, Hannon GJ, Beach D, Lassar AB (1995) Correlation of terminal cell cycle arrest of skeletal muscle with induction of p21 by MyoD. Science 267:1018–1021

Hall M, Bates S, Peters G (1995) Evidence for different modes of action of cyclin-dependent kinase inhibitors: p15 and p16 bind to kinases, p21 and p27 bind to cyclins. Oncogene 11:1581–1588

Harper JW, Adami GR, Wei N, Keyomarsi K, Elledge S J (1993) The p21 Cdk-interacting protein Cip1 is a potent inhibitor of G1 cyclin-dependent kinases. Cell 75:805–816

Harper JW, Elledge SJ, Keyomarsi K et al. (1995) Inhibition of cyclin-dependent kinases by p21. Mol Biol Cell 6:387–400

Hatada I, Mukai T (1995) Genomic imprinting of p57KIP2, a cyclin-dependent kinase inhibitor, in mouse. Nat Genet 11:204–206

Hengst L, Reed SI (1996) Translational control of p27Kip1 accumulation during the cell cycle. Science 271:1861–1864

Hengst L, Dulic V, Slingerland JM, Lees E, Reed SI (1994) A cell cycle-regulated inhibitor of cyclin-dependent kinases. Proc. Natl Acad Sci USA 91:5291–5295

Hirama T, Koeffler HP (1996) Role of cyclin-dependent kinase inhibitors in the development of cancer. Blood 86:841–854

Hollstein M, Sidransky D, Vogelstein B, Harris CC (1991) p53 mutations in human cancers. Science 253:49–53

Kastan MB, Zhan Q, El-Deiry WS, Carrier F, Jacks T, Walsh WV, Plunkett BS, Vogelstein B, Fornace AJ (1992) A mammalian cell cycle checkpoint pathway utilizing p53 and GADD45 is defective in ataxia telangiectasia. Cell 71:587–597

Kato JY, Matsuoka M, Polyak K, Massague J, Sherr CJ (1994) Cyclic AMP-induced G1 phase arrest mediated by an inhibitor (p27Kip1) of cyclin-dependent kinase 4 activation. Cell 79:487–496

Kawamata N, Morosetti R, Miller CW, Park D, Spirin KS, Nakamaki T, Takeuchi S, Hatta Y, Simpson J, Wilcyznski S, et al. (1995) Molecular analysis of the cyclin-dependent kinase inhibitor gene p27/Kip1 in human malignancies. Cancer Res 55:2266–2269

Kawamata N, Seriu T, Koeffler HP, Bartram CR (1996) Molecular analysis of the cyclin-dependent kinase inhibitor family: p16(CDKN2/MTS1/INK4A), p18(INK4C) and p27(Kip1) genes in neuroblastomas. Cancer 77:570–575

Kiwacki RW, Hengst L, Tennant L, Reed SI, Wright PE (1996) Structural studies of p21Waf1/Cip1/Sdi1 in the free and Cdk2 bound state: conformational disorder mediates binding diversity. Proc Natl Acad Sci USA 93:11504–11507

Kiyokawa H, Kineman RD, Manova TK, Soares VC, Hoffman ES, Ono M, Khanam D, Hayday AC, Frohman LA, Koff A (1996) Enhanced growth of mice lacking the cyclin-dependent kinase inhibitor function of p27(Kip1). Cell 85:721–732

Kondo M, Matsuoka S, Uchida K, Osada H, Nagatake M, Takagi K, Harper JW, Takahashi T, Elledge SJ, Takahashi T (1996) Selective maternal-allele loss in human lung cancers of the maternally expressed p57KIP2 gene at 11p15.5. Oncogene 12:1365–1368

Kranenburg O, Scharnhorst V, Van der Eb A, Zantema A (1995) Inhibition of cyclin-dependent kinase activity triggers neuronal differentiation of mouse neuroblastoma cells. J Cell Biol 131:227–234

Lee MH, Reynisdottir I, Massague J (1995) Cloning of p57KIP2, a cyclin-dependent kinase inhibitor with unique domain structure and tissue distribution. Genes Dev 9:639–649

Lee MH, Nikolic M, Baptista CA, Lai E, Tsai LH, Massague J (1996) The brain-specific activator p35 allows Cdk5 to escape inhibition by p27Kip1 in neurons. Proc Natl Acad Sci USA 93:3259–3263

Levine AJ, Momand J, Finlay CA (1991) The p53 tumor suppressor gene. Nature 351:453–456

Li R, Waga S, Hannon GJ, Beach D, Stillman B (1994) Differential effects by the p21 CDK inhibitor on PCNA-dependent DNA replication and repair. Nature 371:534–537

Li R, Hannon GJ, Beach D, Stillman B (1996) Subcellular distribution of p21 and PCNA in normal and repair-deficient cells following DNA damage. Curr Biol 6:189–199

Liu M, Pelling JC (1995) UV-B/A irradiation of mouse keratinocytes results in p53-mediated WAF1/CIP1 expression. Oncogene 10:1955–1960

Liu M, Lee MH, Cohen M, Bommakanti M, Freedman LP (1996) Transcriptional activation of the Cdk inhibitor p21 by vitamin D3 leads to the induced differentiation of the myelomonocytic cell line U937. Genes Dev 10:142–153

Luo Y, Hurwitz J, Massague J (1995) Cell-cycle inhibition by independent CDK and PCNA binding domains in p21Cip1. Nature 375:159–161

Macleod KF, Sherry N, Hannon G, Beach D, Tokino T, Kinzler K, Vogelstein B, Jacks T (1995) p53-dependent and independent expression of p21 during cell growth, differentiation, and DNA damage. Genes Dev 9:935–944

Mal A, Poon R, Howe PH, Toyoshima H, Hunter T, Harter ML (1996) Inactivation of $p27^{Kip1}$ by the viral E1A oncoprotein in TGF-β treated cells. Nature 380:262–265

Matsuoka S, Edwards MC, Bai C, Parker S, Zhang P, Baldini A, Harper JW, Elledge SJ (1995) p57KIP2, a structurally distinct member of the p21CIP1 Cdk inhibitor family, is a candidate tumor suppressor gene. Genes Dev 9:650–662

Matsuoka S, Thompson JS, Edwards MC, Barletta JM, Grundy P, Kalikin LM, Harper JW, Elledge SJ, Feinberg AP (1996) Imprinting of the gene encoding a human cyclin-dependent kinase inhibitor, $p57^{Kip2}$, on chromosome 11p15. Proc Natl Acad Sci USA 93:3026–3030

McDonald ER, Wu GS, Waldman T, El Deiry WS (1996) Repair defect in p21 WAF1/CIP1-/- human cancer cells. Cancer Res 56:2250–2255

Medrano EE, Im S, Yang F, Abdel MZ (1995) Ultraviolet B light induces G1 arrest in human melanocytes by prolonged inhibition of retinoblastoma protein phosphorylation associated with long-term expression of the p21Waf-1/SDI-1/Cip-1 protein. Cancer Res 55:4047–4052

Nakanishi M, Robetorye RS, Adami GR, Pereira SO, Smith JR (1995a) Identification of the active region of the DNA synthesis inhibitory gene p21Sdi1/CIP1/WAF1. Embo J 14:555–563

Nakanishi M, Robetorye RS, Pereira SO, Smith JR (1995b) The C-terminal region of p21SDI1/WAF1/CIP1 is involved in proliferating cell nuclear antigen binding but does not appear to be required for growth inhibition. J Biol Chem 270:17060–17063

Nakayama K, Ishida N, Shirane M, Inomata A, Inoue T, Shishido N, Horii I, Loh DY, Nakayama K (1996) Mice lacking p27(Kip1) display increased body size, multiple organ hyperplasia, retinal dysplasia, and pituitary tumors. Cell 85:707–720

Noda A, Ning Y, Venable SF, Pereira SO, Smith JR (1994) Cloning of senescent cell-derived inhibitors of DNA synthesis using an expression screen. Exp Cell Res 211:90–98

Nourse J, Firpo E, Flanagan WM, Coats S, Polyak K, Lee MH, Massague J, Crabtree GR, Roberts JM (1994) Interleukin-2-mediated elimination of the p27Kip1 cyclin-dependent kinase inhibitor prevented by rapamycin. Nature 372:570–573

Orlow I, Iavarone A, Crider-Miller SJ, Bonilla F, Latres E, Lee M-H, Gerald WL, Massague J, Weissman BE, Cordon-Cardo C (1996) Cyclin-dependent kinase inhibitor, $p57^{Kip2}$ in soft tissue sarcomas and Wilm's tumors. Cancer Res 56:1219–1221

Pagano M, Tam SW, Theodoras AM, Beer-Romero P, Del Sal G, Chau V, Yew PR, Draetta GF, Rolfe M (1995) Role of the ubiquitin-proteasome pathway in regulating abundance of the cyclin-dependent kinase inhibitor p27. Science 269:682–685

Pan ZQ, Reardon JT, Li L, Flores RH, Legerski R, Sancar A, Hurwitz J (1995) Inhibition of nucleotide excision repair by the cyclin-dependent kinase inhibitor p21. J Biol Chem 270:22008–22016

Parker SB, Eichele G, Zhang P, Rawls A, Sands AT, Bradley A, Olson EN, Harper JW, Elledge SJ (1995) p53-independent expression of p21Cip1 in muscle and other terminally differentiating cells. Science 267:1024–1027

Petrocelli T, Poon R, Drucker DJ, Slingerland JM, Rosen CF (1996) UVB radiation induces p21Cip1/WAF1 and mediates G1 and S phase checkpoints. Oncogene 12:1387–1396

Pietenpol JA, Bohlander SK, Sato Y, Papadopoulos N, Liu B, Friedman C, Trask BJ, Roberts JM, Kinzler KW, Rowley JD et al. (1995) Assignment of the human p27Kip1 gene to 12p13 and its analysis in leukemias. Cancer Res 55:1206–1210

Podust VN, Podust LM, Goubin F, Ducommun B, Hubscher U (1995) Mechanism of inhibition of proliferating cell nuclear antigen-dependent DNA synthesis by the cyclin-dependent kinase inhibitor p21. Biochemistry 34:8869–8875

Pollock PM, Pearson JV, Hayward NK (1996) Compilation of somatic mutations of the CDKN2 gene in human cancers:non-random distribution of base substitutions. Genes Chrom Cancer 15:77–88

Polyak K, Kato JY, Solomon MJ, Sherr CJ, Massague J, Roberts JM, Koff A (1994a) p27Kip1, a cyclin-Cdk inhibitor, links transforming growth factor-beta and contact inhibition to cell cycle arrest. Genes Dev 8:9–22

Polyak K, Lee MH, Erdjument BH, Koff A, Roberts JM, Tempst P, Massague J (1994b) Cloning of p27Kip1, a cyclin-dependent kinase inhibitor and a potential mediator of extracellular antimitogenic signals. Cell 78:59–66

Ponce CM, Lee MH, Latres E, Polyak K, Lacombe L, Montgomery K, Mathew S, Krauter K, Sheinfeld J, Massague J, Cordon-Cardo C (1995) p27Kip1: chromosomal mapping to 12p12–12p13.1 and absence of mutations in human tumors. Cancer Res 55:1211–1214

Poon RY, Toyoshima H, Hunter T (1995) Redistribution of the CDK inhibitor p27 between different cyclin-CDK complexes in the mouse fibroblast cell cycle and in cells arrested with lovastatin or ultraviolet irradiation. Mol Biol Cell 6:1197–1213

Reid LH, Crider-Miller SJ, West A, Lee M.-H, Massague J, Weissman BE (1996) Genomic organization of human p57Kip2 gene and its analysis in the G401 Wilm's tumor assay. Cancer Res 56:1214–1218

Resnitzky D, Hengst L, Reed SI (1995) Cyclin A-associated kinase complex activity is rate limiting for entrance into S phase and is negatively regulated in G1 by $p27^{Kip1}$. Mol Cell Biol 15:4347–4352

Reynisdottir I, Polyak K, Iavarone A, Massague J (1995) Kip/Cip and Ink4 Cdk inhibitors cooperate to induce cell cycle arrest in response to TGF-beta. Genes Dev 9:1831–1845

Russo AA, Jeffrey PD, Patten AK, Massague J, Pavletich NP (1996) Crystal structure of the $p27^{Kip1}$ cyclin-dependent-kinase inhibitor bound to the cyclin A-Cdk2 complex. Nature 382:325–331

Schulze A, Zerfass-Thome K, Berges J, Middendorf S, Jansen-Duerr P, Henglein B (1996) Anchorage-dependent transcription of the cyclin A gene. Mol Cell Biology 16:4632–4638

Serrano M, Lee H-W, Chin L, Cordon-Cardo C, Beach D, DePinho RA (1996) Role of the INK4a locus in tumor suppression and cell mortality. Cell 85:27–37

Shiohara, M, El-Deiry WS, Wada M, Nakamaki T, Takeuchi S, Yang R, Chen DL, Vogelstein B, Koeffler HP (1994) Absence of WAF1 mutations in a variety of human malignancies. Blood 84:3781–3784

Shivji MK, Grey SJ, Strausfeld UP, Wood RD, Blow JJ (1994) Cip1 inhibits DNA replication but not PCNA-dependent nucleotide excision-repair. Current Biology 4:1062–1068

Shou W, Dunphy WG (1996) Cell cycle control by Xenopus $p28^{Kix1}$, a developmentally regulated inhibitor of cyclin dependent kinases. Mol Biol Cell 7:457–469

Skapek SX, Rhee J, Spicer DB, Lassar AB (1995) Inhibition of myogenic differentiation in proliferating myoblasts by cyclin D1-dependent kinase. Science 267:1022–1024

Slingerland JM, Hengst L, Pan CH, Alexander D, Stampfer MR, Reed SI (1994) A novel inhibitor of cyclin-Cdk activity detected in transforming growth factor beta-arrested epithelial cells. Mol Cell Biol 14:3683–3694

Soos TJ, Kiyokawa H, Yan JS, Rubin MS, Giordano A, DeBlasio A, Bottega S, Wong B, Mendelsohn J, Koff A (1996) Formation of p27-Cdk complexes during the human mitotic cell cycle. Cell Growth Diff 7:135–146

Su JY, Rempel RE, Erikson E, Maller JL (1995) Cloning and characterization of the Xenopus cyclin-dependent kinase inhibitor $p27^{XIC1}$. Proc Natl Acad Sci USA 92:10187–10191

Sun Y, Hildesheim A, Li H, Li Y, Chen JY, Cheng YJ, Hayes RB, Rothman N, Bi WF, Cao Y et al. (1995) No point mutation but a codon 31ser→arg polymorphism of the WAF-1/CIP-1/p21 tumor suppressor gene in nasopharyngeal carcinoma (NPC): the polymorphism distinguishes Caucasians from Chinese. Cancer Epidemiol Biomarkers Prev 4:261–267

Takeuchi S, Mori N, Koike M, Slater J, Park S, Miller CW, Miyoshi I, Koeffler, HP (1996) Frequent loss of heterozygosity in region of the KIP1 locus in non-small cell lung cancer: evidence for a new tumor suppressor gene on the short arm of chromosome 12. Cancer Res 56:738–740

Toyoshima H, Hunter T (1994) p27, a novel inhibitor of G1 cyclin-Cdk protein kinase activity, is related to p21. Cell 78:67–74

Vidal MJ, Loganzo FJ, de Oliveira AR, Hayward NK, Albino AP (1995) Mutations and defective expression of the WAF1 p21 tumour-suppressor gene in malignant melanomas. Melanoma Res 5:243–250

Waga S, Hannon GJ, Beach D, Stillman B (1994) The p21 inhibitor of cyclin-dependent kinases controls DNA replication by interaction with PCNA. Nature 369:574–578

Waldman T, Kinzler KW, Vogelstein B (1995) p21 is necessary for the p53-mediated G1 arrest in human cancer cells. Cancer Res 55:5187–5190

Wang QM, Jones JB, Studzinski GP (1996) Cyclin-dependent kinase inhibitor p27 as a mediator of the G1-S phase block induced by 1,25-dihydroxyvitamin D_3 in HL60 cells. Cancer Res 56:264–267

Warbrick E, Lane DP, Glover DM, Cox LS (1995) A small peptide inhibitor of DNA replication defines the site of interaction between the cyclin-dependent kinase inhibitor p21WAF1 and proliferating cell nuclear antigen. Curr Biol 5:275–282

Watanabe H, Fukuchi K, Takagi Y, Tomoyasu S, Tsuruoka N, Gomi K (1995) Molecular analysis of the Cip1/Waf1 (p21) gene in diverse types of human tumors. Biochim Biophys Acta 1263:275–280

Winston J, Dong F, Pledger WJ (1996) Differential modulation of G1 cyclins and the Cdk inhibitor p27kip1 by platelet-derived growth factor and plasma factors in density-arrested fibroblasts. J Biol Chem 271:11253–11260

Xiong Y, Zhang H, Beach D (1992) D type cyclins associate with multiple protein kinases and the DNA replication and repair factor PCNA. Cell 71:505–514

Xiong Y, Hannon GJ, Zhang H, Casso D, Kobayashi R, Beach D (1993a) p21 is a universal inhibitor of cyclin kinases. Nature 366:701–704

Xiong Y, Zhang H, Beach D (1993b) Subunit rearrangement of the cyclin-dependent kinases is associated with cellular transformation. Genes Dev 7:1572–1583

Zhang H, Xiong Y, Beach, D. (1993) Proliferating cell nuclear antigen and p21 are components of multiple cell cycle kinase complexes. Mol Biol Cell 4:897–906

Zhang H, Hannon GJ, Beach D (1994a) p21-containing cyclin kinases exist in both active and inactive states. Genes Dev 8:1750–1758

Zhang H, Hannon GJ, Casso D, Beach D (1994b) p21 is a component of active cell cycle kinases. Cold Spring Harb Symp Quant Biol 59:21–29

Zhu X, Ohtsubo M, Bohmer RM, Roberts JM, Assoian RK (1996) Adhesion-dependent cell cycle progression linked to the expression of cyclin D1, activation of cyclin E-cdk2, and phosphorylation of the retinoblastoma protein. J Cell Biol 133:391–403

The INK4 Family of CDK Inhibitors

A. Carnero[1,2] and G.J. Hannon[1]

1 Introduction 43
2 p16INK4a 45
3 p15INK4b 49
4 p18INK4c and p19 INK4d 50
References 52

1 Introduction

Decisions concerning the fate of a cell are intimately linked to the proliferative state of that cell. Proliferation of certain cell populations is required to maintain or repair tissues in an aging organism. However, this proliferation must be tightly regulated. Failure to control proliferation may interfere with differentiation, causing a cell to fail to achieve its fully determined state. Potentially more severe consequences could also ensue from uncontrolled cell division. When this is accompanied by a failure of programmed cell death, a cancerous tumor can result.

The proliferative state of a cell is ultimately translated into control of a group of related protein kinases known as the cyclin-dependent kinases (CDKs). These, in partnership with essential positive regulatory subunits, cyclins, control progress through the cell division cycle by phosphorylation of key substrates. Although CDKs are clearly central, other mechanisms control cell cycle pathways that operate independently of the nuclear division cycle (Lu and Hunter 1995).

The cell cycle can be operationally divided into two parts, one in which the decision to proliferate is made and another in which the decision to proliferate is executed. During the decisive portion of the cycle, information concerning the extracellular environment and intracellular state of a cell is integrated through a number of regulatory pathways which can cause a cell either to cease growth or to enter the division cycle. Once a cell has entered the division cycle, it generally

---

[1] Cold Spring Harbor Laboratory, P.O. BOX 100, Cold Spring Harbor, NY 11724, USA
[2] Present address: Institute of Child Health, 30 Guilford Street, London WC1N 1EH, UK

becomes insensitive to extracellular signals but can still arrest response to intracellular checkpoints which assess the integrity of the division process.

It has long been known that the decision to proliferate is made during the G1 phase of the division cycle. This choice is controlled by two classes of cyclin/CDK enzymes: cyclin D/CDK4(6) and cyclin E/CDK2. Of these, the cyclin D/CDK4 enzyme acts first and is generally thought to be the key downstream recipient of positive and negative extracellular signals (see SHERR 1993). However, mounting evidence suggests that some extracellular signals (such as those which insure appropriate cell–matrix contacts) may operate through the regulation of cyclin E/CDK2 (FANG et al. 1996; ZHU et al. 1996).

Although cyclin D/CDK4 may modify numerous substrates, only one is critical to the ability of the enzyme to control proliferation of cells in culture (SHERR 1993, 1994). This is the product of the retinoblastoma susceptibility gene (pRb). Throughout early G1 phase, Rb exists in a hypophosphorylated state that inhibits the entry of a cell into the division cycle (see WEINBERG 1995). Rb may accomplish this through a variety of mechanisms but prominent among these is inhibition of the expression of gene products required for the execution of subsequent cell cycle phases. Phosphorylation of Rb and its conversion into a non-inhibitory form requires both the cyclin D/CDK4 and cyclin E/CDK2 enzymes (WEINBERG 1995).

Cells that lack Rb do not require the activity of cyclin D-associated kinases for growth, suggesting that Rb is the only critical substrate for these enzymes in cells in tissue culture (SHERR and ROBERTS 1995). In vivo, the situation may be more complex. Rb has two closely related family members, p107 and p130, which have properties that are grossly similar to those of Rb (WEINBERG 1995). It is as yet unclear whether these relatives may be critical substrates for CDKs in some cell types. Cells that lack Rb do, however, still require the activity of cyclin E/CDK2 for growth (SHERR 1993, 1994). Thus the role of the cyclin E-associated enzyme is clearly more broad than that of the cyclin D-containing complexes.

The activity of the cyclin D-CDK4 and cyclin D-CDK6 enzymes is tightly controlled by a number of regulatory mechanisms. First, the activity of the kinase is regulated by the availability of the cyclin subunit. CDK4 and the closely related kinase, CDK6, associate exclusively with D-type cyclins (D1, D2, and D3). Expression of D-cyclin is not obviously cell cycle-dependent but is instead controlled primarily by extracellular signals (SHERR 1993). Once cyclin D and CDK4 subunits are available, their association appears to require an assembly factor whose activity may also respond to extracellular growth stimuli (MATSUSHIME et al. 1994). Once assembled, complexes are subject to both the necessity for activating phosphorylation by CDK-activating kinase (CAK) and potential inhibitory phosphorylation of a tyrosine residue in the ATP binding site (DRAETTA 1990). While CAK phosphorylation appears to be constitutive, inhibitory phosphorylation may constitute a regulatory mechanism (as has been amply demonstrated for other CDK enzymes) (KING et al. 1994). Finally, the cyclin D-CDK4 enzyme can be regulated by association with stoichiometric inhibitors of either of two families: the CIP/KIP family or the INK4 family.

The CIP/KIP family are general CDK inhibitors that are discussed in detail elsewhere in this volume. The focus of this review is the INK4 family of CDK regulators. The INK4 family consists currently of four related proteins, p16 INK4a, p15 INK4b, p18 INK4c, and p19 INK4d (SERRANO et al. 1993; HANNON and BEACH 1994; CHAN et al. 1995; HIRAI et al. 1995; GUAN et al. 1994). These four proteins specifically bind to and inactivate CDK4 and CDK6 kinases. Although these four proteins are indistinguishable at the biochemical level, the more detailed discussion of each in the following sections should indicate that each member of the INK4 family is used by a cell to control proliferative choices under specific circumstances.

2 p16INK4a

The p16 protein was first noted in a study that was targeted at identifying differences in the cell cycle regulators of normal and transformed cells. In SV40-transformed fibroblasts, the G1-regulatory kinase, CDK4, was not found in association with cyclin D, but was instead bound to a protein with an apparent M_r of 16 kDa. The gene encoding p16 was cloned by a two-hybrid approach in which CDK4 was used as the interaction target. Biochemical experiments suggested that p16 was an inhibitor of CDK4 kinase, and it was thus dubbed p16 INK4 (inhibitor of CDK4). This was an unexpected result since p16 was complexed with a cell cycle regulatory kinase in highly proliferative, transformed cells. However, this paradox was resolved by the fact that cells expressing SV40 T-antigen no longer require the activity of CDK4 for cell cycle progression.

The p16 gene encodes a polypeptide of 167 amino acids in mouse and 156 amino acids in humans (QUELLE et al. 1995a). The most striking feature of the p16 sequence is that it is composed almost entirely of four ankyrin repeat units, suggesting that the protein might be folded from helix-β-turn-helix motifs (KALUS et al. 1997). p16 proteins bind specifically to CDK4 and CDK6 with a 1:1 stoichiometry (SERRANO et al. 1993; HANNON et al. 1994). The result of this binding is loss of CDK4 kinase activity toward its physiological substrate, Rb. p16 binds CDK4 and CDK6 in the absence of cyclin D, and purified p16 protein can promote dissociation of the CDK4-cyclin D1 complex (HALL et al. 1995; SERRANO et al. 1993). It is as yet unclear whether displacement of the cyclin is the primary mechanism by which p16 inhibits CDK4 and CDK6 kinases, or whether destabilization of the cyclin–CDK interaction is a secondary consequence of allosteric changes in the structure of the CDK enzyme. Although the ankyrin repeat motif is the predominant structural feature of p16, not all proteins containing ankyrin motifs can effectively inhibit CDK4/6 enzymes. For instance, mouse inhibitor of NF-κB (IκB) or a subdomain of the protein containing only ankyrin repeats is a much less efficient inhibitor of CDK4/6 than is p16 (HIRAI et al. 1995).

Cyclin D, but not cyclin A or cyclin E, can bind directly to pRb, targeting CDK4/6 to its substrate (SHERR 1994; PINES 1996). Hypophosphorylated Rb binds to E2F-DP1 heterodimers, preventing this complex from activating the transcription of genes needed for the transition from G1 into S-phase (WEINBERG 1995). Phosphorylation of pRb by CDK4/6 and by cyclin E/CDK2 negates the growth-suppressive effects of Rb via release of E2F and consequent activation of genes required for DNA replication. As predicted by the forgoing model, ectopic expression of p16 in cell lines that have an intact Rb gene arrests cells in G1 and prevents growth. However, pRb-deficient cells are insensitive to the inhibitory effects of p16 overexpression (GUAN et al. 1994; SERRANO et al. 1995; MEDEMA et al. 1995; LUKAS et al. 1995). These findings are consistent with the observation that the microinjection of anti-cyclin D1 antibodies or cyclin D1 antisense plasmids causes G1 arrest in normal fibroblasts but has no effect in cells lacking functional pRb (TAM et al. 1994).

Several lines of evidence suggested that p16 might be a tumor suppressor. First, as an inhibitor of CDK4, p16 is also a growth suppressor, and the key target of CDK4, Rb, is a tumor suppressor in its own right (WEINBERG 1995). Thus, p16 is a negative regulator of a protein that functions to inactivate a tumor/growth suppressor in proliferating cells. Also, cyclin D, an activator of CDK4, is a demonstrated oncogene (SHERR 1993, 1994; HUNTER and PINES 1994). The hypothesis that p16 acts as a tumor suppressor was supported by isolation of the gene encoding p16 through a cytogenetic approach which was designed to identify the tumor suppressor that had been mapped to 9p21 (KAMB et al. 1994). Following this revelation, a great deal of effort was directed at determining the relevance of p16 inactivation to tumors. This is discussed in detail elsewhere in this volume; however, it is of use to summarize some of the information here since it was largely information from tumors and tumor cells that led to the first insights into the biological function of p16.

The p16 gene is altered in a high percentage of human tumors of many different varieties. p16 can be inactivated by a variety of mechanisms including deletion, point mutation, and silencing by methylation (see the chapter by A. KAMB, this volume). However, the final confirmation that p16 was indeed a tumor suppressor came from the generation of p16-null mice. p16 INK4a proved non-essential for viability or for proper development (SERRANO et al. 1996). INK4a null mice did show some abnormal features that might indicate a role for p16 in the development of some specific cell types. For example, p16-null mice have features consistent with abnormal extramedullary hematapoeisis, suggesting that p16 might regulate the proliferation of some hematapoeitic lineages or their progenitors. This role in hematopoeisis is not unexpected since pRb null mice are severely impaired in the production of mature erythrocytes and have increased numbers of megakaryocytes and myeloid cells in liver (LEE et al. 1992; JACKS et al. 1992; CLARKE et al. 1992). As predicted by studies of human tumors, INK4a null mice develop spontaneous tumors (mostly soft tissue sarcomas and B-cell lymphomas) at an accelerated rate. Furthermore, p16-null mice are highly susceptible to the development of tumors following carcinogenic treatments [dimethylbenzanthracene/ultraviolet (DMBA/UV) or DMBA alone] (SERRANO et al. 1996).

Interpretation of the results from p16-null mice was complicated by the discovery of a second growth regulatory protein that is also encoded from within the p16 locus (QUELLE et al. 1995b). p19 ARF is encoded by an alternatively spliced mRNA in which an alternative first exon (E1b) is appended to exon 2 of the p16 mRNA. The protein from this mRNA is translated from an alternative reading frame and thus shows no homology at the protein level with the INK4 family of cyclin-dependent kinase inhibitors. p19 ARF has the ability to induce growth arrest upon enforced expression; however, in contrast to p16, p19 arrested cells distribute between G1 and G2 phases. Although the mechanism by which p19 regulates cell proliferation is not known, it does not seem to involve direct inhibition of cyclin/CDK enzymes.

The question of whether tumor suppression is mediated by p16 INK4, by p19 ARF, or by a combination of both is settled by an examination of the point mutations that are found in human tumors. Altered p16 proteins that contain point mutations are impaired in their ability to bind and to inhibit the CDK4 enzyme (KOH et al. 1995). A number of these mutations do not affect the reading frame which gives rise to the ARF protein. Further confirmation that tumor suppression by the p16 locus occurs via p16 INK4 comes from the isolation of a mutant CDK4 from a number of melanomas (WOLFEL et al. 1995). This allele is competent in its ability to bind cyclin D and to phosphorylate Rb; however, the CDK4 mutant shows a greatly reduced sensitivity to inhibition by p16 INK4. The CDK4 mutant, however, is still sensitive to inhibition by the CIP/KIP family of proteins (WOLFEL et al. 1995).

The fact that p16 was highly expressed in some tumors was the first hint towards the mechanism by which p16 might function as a tumor suppressor. Initially, p16 overexpression in Rb-negative tumor cells suggested a feedback loop in which Rb might act as a suppressor of p16 expression, much as it acted to suppress the expression of S-phase specific genes during G1. However, two observations argued against this model. Firstly, expression of p16 was not obviously cell-cycle dependent (HARA et al. 1996). Secondly, mouse embryo fibroblasts from animals that lacked Rb genes did not show an obviously elevated p16 level (MEDEMA et al. 1995). These counter-indications spawned the hypothesis that p16 induction might occur as a consequence of some aspect of the transformation process itself. This hypothesis was supported by observations made with fibroblasts derived from p16-null embryos.

At the cellular level, p16/p19ARF-deficient cells displayed altered growth properties. p16/p19ARF-deficient MEFs proliferate rapidly and show a high colony formation efficiency when plated at low density (SERRANO et al. 1996). Furthermore, cultures of p16-null cells did not display a lag in growth characteristic of the M1 senescence control in primary mouse cells. Finally, p16 abundance increased during senescence in human cells. Considered together, these results suggested that p16 might play a role in creating the senescent phenotype. In this model, cells lacking p16 might acquire an extended life span that could promote immortalization. By extension, p16-null embryo fibroblasts might be predisposed to immortalization or be essentially immortal without further mutation.

The activated *ras* oncogene is a potent transforming stimulus in immortal cells. Its effect on normal cells, however, is quite different. Expression of activated *ras* in primary human or mouse fibroblasts induces growth arrest (SERRANO et al. 1997). This arrest is indistinguishable from cellular senescence based upon a number of well-established senescence markers. In contrast, introduction of oncogenic *ras* into primary embryo fibroblasts derived from p16-null mice causes transformation. A similar response is seen in p53-null cells.

ras-induced senescence of normal cells suggests that these cells possess a "gatekeeper" mechanism which senses inappropriate mitogenic stimuli. This mechanism may then trigger cellular pathways that irreversibly withdraw that cell from the division cycle and place the cell in a senescent state. Accordingly, activation of *ras* should provoke activation of the p53 and p16 pathways since inactivation of these pathways correlated with loss of *ras*-induced senescence. Indeed, upon introduction of activated *ras* into normal human or mouse fibroblast cells, p16 levels increased (SERRANO et al. 1997). Elevation of p21, a p53 target that mediates growth arrest, has also been observed.

The foregoing observations lead to a model in which oncogenic stimuli, such as constitutive *ras* activation, trigger a p16 response which arrests cell proliferation by preventing Rb phosphorylation. Transformed cells may overcome this block through a number of mechanisms. In the majority of cases, p16 is inactivated by deletion, mutation, or methylation. In other cases, Rb function may be lost by deletion, mutation or by expression of a viral oncoprotein. Cells may also increase expression of cyclin D which could compete with p16 for binding to CDK4 kinase. Finally, several cases have been reported in which tumor cells posses CDK4 alleles that resist inhibition by INK4 family members (WOLFEL et al. 1995). Although activation of the *ras* pathway is common in human tumors, the paramount importance of the p16 pathway in tumor suppression predicts that this pathway may also respond to a broader range of oncogenic stimuli.

Although the role of p16 in tumor suppression is clear, p16 may not be limited to this function. Induction of p16 has been observed during terminal differentiation of a teratocarcinoma cell line, NT2, into post-mitotic neurons and during the transition from fetal to adult brain (LOIS et al. 1995). Some oncogenes, particularly *ras*, can provoke terminal differentiation in some cell lines. Enforced ras expression promotes conversion of PC12 rat pheochromocytoma to a "neural" phenotype (BAR-SAGI and FERAMISCO 1985) and causes differentiation of 3T3-L1 fibroblasts into adipocytes (BENITO et al. 1991). The signal transduction pathways activated by oncogenic *ras* during the differentiation process clearly resemble the *ras*-induced pathway leading to fibroblast proliferation. It is therefore possible that *ras*-provoked induction of p16 is also a feature of the differentiation process, contributing to the withdrawal of differentiated cells from the division cycle. This is consistent with the fact that p16 expression can cause differentiation of myoblasts into myocytes (SKAPEK et al. 1995; WANG et al. 1996). In cells that are not appropriately primed to differentiate, p16 expression may have the default effect of pushing cells down the senescence pathway. However, circumstantial evidence for the role of p16 in terminal differentiation is countered by the fact that mice lacking p16 are substantially normal (SERRANO et al. 1996). Thus, any role of p16 in this process must be redundant with other pathways.

3 p15 INK4b

The gene encoding p15 INK4b was first noted by KAMB et al. during their cytogenetically-based cloning of p16 INK4a. While the coding sequence of p16 was clearly altered in tumor cell lines, a closely linked genomic segment that showed high homology to p16 exon 2 was unaltered. Furthermore, this orphan exon was not linked to a genomic segment with obvious homology to p16 exon 1. Thus, the manuscript that described the gene encoding p16 as the tumor suppressor at 9p21 implied that the orphan exon 2 homolog was a probable pseudogene (KAMB et al. 1994). This later proved untrue as other groups demonstrated that a close p16 relative, p15 INK4b, was also encoded in the 9p21 locus (HANNON and BEACH 1994; JEN et al. 1994).

The p15 INK4b cDNA was isolated following a series of experiments designed to identify the point at which a growth inhibitory cytokine, transforming growth factor (TGF)-β, impacted the cell-cycle regulatory machinery (HANNON and BEACH 1994). Treatment of a human keratinocyte cell line, HACAT, with TGF-β causes cell cycle arrest in the G1 phase. This is accompanied by a rearrangement of the multiprotein complexes that contain the G1-regulatory kinases, CDK4 and CDK6. Upon TGF-β treatment, a small (approximately 15 kDa) protein became associated with CDK4 and CDK6 concomitant with loss of CDK4 and CDK6 activity. This protein was initially thought to be p16; however, V8 protease mapping demonstrated that p15 was distinct. p15 was, however, recognized weakly by the p16 antiserum prompting the cloning of p15 via a low-stringency hybridization approach.

As it happened, low stringency was unnecessary as p15 and p16 share a high degree of identity in their C-terminal segments. The N-terminal portions of these proteins are much more diverged; however, the four-ankyrin repeat structure that is characteristic of the INK4 family is maintained. p15 and p16 are biochemically indistinguishable in vitro. Both bind to and inhibit CDK4 and CDK6 kinases with similar affinities. The genes encoding these proteins are closely linked on chromosome 9 at p21, suggesting that these homologs arose from a gene duplication event. This event preceded divergence of mice and humans since the genomic organization of p15 and p16 is preserved in humans and rodents.

Despite their close linkage and functional conservation at the biochemical level, p15 and p16 play vastly different biological roles. The abundance of p15 is dramatically altered upon treatment of a variety of different cells with TGF-β. Such cells include mouse and human keratinocytes, mink lung, and human mammary epithelial cells (HANNON and BEACH 1994; REYNISDOTTIR et al. 1995). Induction of p15 occurs at both the post-transcriptional and transcriptional levels, and TGF-β responsiveness has been mapped in the p15 promoter to an SP1 element (LI et al. 1995). TGF-β is a multi-functional cytokine that can act as a growth factor, a differentiation factor, or a growth inhibitor (see MASSAGUE and POLYAK 1995). In all cell types in which TGF-β has been shown to inhibit growth, TGF-β treatment also causes induction of p15. Following induction, p15 accumulates in CDK4 and CDK6 complexes and renders them catalytically inactive (SANDHU et al. 1997).

Because of the requirement for CDK4 and CDK6 kinase activity during the G1 phase, induction of p15 and consequent inhibition of CDKs is sufficient to explain growth inhibition by TGF-β. However, recent data suggest that the situation may be more complex. p15 was not the only CDK inhibitor that was identified based upon a connection to TGF-β-mediated growth inhibition. p27KIP1, a member of the p21 family of CDK inhibitors, was purified based upon its increased activity in lysates from TGF-β-arrested cells (POLYAK et al. 1994). Although the abundance of p27 protein is not altered in response to TGF-β, the proteins with which p27 associates are changed (REYNISDOTTIR et al. 1995). Upon TGF-β treatment, p27 appears to relocate from cyclin D-CDK4/6 complexes to cyclin E-CDK2 complexes. Studies both in vitro and in vivo support the idea that this alteration probably occurs as a consequence of p27 displacement by p15 INK4b (REYNISDOTTIR and MASSAGUE 1997). In this way, the transcriptional activation of a CDK4/6 specific inhibitor can achieve inhibition of both CDK4/6 and CDK2 kinases.

The displacement model seemed to provide a coherent explanation for the ability of TGF-β to enforce growth arrest by regulation of two critical classes of cyclin-dependent kinases. However, a cell line lacking the p15 gene can still arrest following treatment with TGF-β (IAVARONE and MASSAGUE 1997). This points to the existence of additional, possibly redundant, pathways by which TGF-β can regulate proliferation. One candidate for such a pathway is the loss of cdc25 protein that can occur following TGF-β treatment (IAVARONE and MASSAGUE 1997) as a downstream consequence of loss of *myc* expression (GALAKTIONOV et al. 1996).

Despite the plethora of possibilities, no one has provided a definitive model for the mechanism by which TGF-β causes growth arrest. This effort is complicated by the redundancy that is rife in mammalian growth control pathways and also by the fact that TGF-β negatively affects the growth of many different cell types. Further muddling the situation is the fact that TGF-β can also promote either growth or differentiation. Thus, it is possible that individual targets (e.g., p15) might be critical in some cell types, whereas in others alternative mechanisms (e.g., *myc*) may predominate. This is of interest in light of the fact that loss of TGF-β sensitivity is a common characteristic of human tumor cells and that loss of the p15 gene is also a common cytogenetic event in transformed cells. While this may obviously occur as a bystander-effect of deletion of p16, preferential loss of the p15 gene is observed in a limited, but coherent, subset of tumor types (JEN et al. 1994; ZHOU and LINDER 1996; KAWAMATA et al. 1995; ZHANG et al. 1996; ZARIWALA et al. 1996).

4 p18INK4c and p19 INK4d

The p18 and p19 genes were first isolated in yeast two-hybrid screens that were designed to search for proteins that interact with CDK4 (HIRAI et al. 1995) or

CDK6 (GUAN et al. 1994). p19 was also cloned through its interaction in yeast with Nur 77, an orphan steroid receptor that shows no link to cell cycle control (CHAN et al. 1995). p18 and p19 are approximately 40% identical and have similar degrees of protein homology with p16 and p15. As is characteristic of the INK4 family, p18 and p19 are composed of four tandem ankyrin motifs, each about 32 amino acids in length.

p18 and p19 specifically bind to the cyclin D-dependent catalytic subunits, CDK4 and CDK6, and are unable to interact with other CDKs or directly with D-cyclins (HIRAI et al. 1995). In vitro, GST-p18 and GST-p19 bind with the same affinity to CDK4 and CDK6 (HIRAI et al. 1995). However, in immunoprecipitations of CDK complexes, both p18 and p19 show a preferential association with CDK6 (HIRAI et al. 1995; GUAN et al. 1994). In contrast with other members of the INK4, p18 and p19 do not displace cyclin D from CDK4 and CDK6. In vitro reconstitution of the complexes demonstrates that recombinant p18 and p19 proteins can bind to CDK4–cyclin D complexes without displacing the cyclin (HIRAI et al. 1995). However, in vivo, p18 or p19 do not co-immunoprecipitate with cyclin D. This suggests that p18 and p19 also prevent the binding of cyclin D to the catalytic subunit in vivo.

As with p15 and p16, the result of p18 or p19 binding to CDKs or CDK–cyclin D complexes is the inhibition of CDK. However, none of these proteins inhibit the pRb kinase activity of cyclin E–CDK2 complexes (HIRAI et al. 1995). Despite evidence of inhibition of CDK4/6 by cyclin D displacement, another mechanism for the inhibition of CDKs by p18 has also been proposed. p18 can efficiently block CDK6 phosphorylation by CAK but has no effect on the phosphorylation and activation of CDK2 (APRELIKOVA et al. 1995). Since p18 can not bind directly to CDK7 or cyclin H (CAK subunits), it has been suggested that p18 blocks CDK6 activation by rendering this protein inaccessible to CAK.

As predicted from the forgoing biochemical experiments, overexpression of p18 or p19 in cells that contain functional pRb significantly reduces the ability of these cells to form colonies in tissue culture (GUAN et al. 1994; HIRAI et al. 1995). However, overexpression of p18 has no effect on the growth of cells that lack Rb (GUAN et al. 1994).

p19 gene is ubiquitously expressed as a single 1.4-kb mRNA transcript with the highest levels in thymus, spleen, peripheral blood leukocytes, fetal liver, brain, and testes (HIRAI et al. 1995; CHAN et al. 1995; OKUDA et al. 1995). In mouse fibroblasts and macrophages, p19 is expressed at very low levels during the G1 phase, but is rapidly induced at the G1/S transition and remains at high levels during the remainder of the cell cycle (OKUDA et al. 1995).

p18 expression patterns are more complex. There are at least three different p18 transcripts of 2.5, 1.9, and 1.1 kb. These are differentially expressed in various tissues (HIRAI et al. 1995). During entry of cells into the cell cycle, p18 mRNA is induced with kinetics similar to those of p19. CSF-1 starved macrophages expressed the 1.1-kb transcript, and only the 1.9-kb transcript is induced at the G1/S transition (HIRAI et al. 1995). In Daudi cells (a hematapoeitic cell line) the ratio between the 2.5-kb and the 1.9-kb transcripts are changed after treatment with a growth inhibitory cytokine, interferon (IFN)α. Treatment provoked a decrease in the 2.5-kb transcript and an increase in the 1.9-kb transcript (SANGFELT et al. 1997). However, the consequences of these changes are unknown and no variations in the levels of the p18 transcripts are observed after IFNα treatment of other hematopoietic cell lines. IFNα does not affect the levels of p19 in these cell lines; however, treatment of the murine myeloid leukemia cell line, M1, with interleukin (IL)-6 specifically induces p19. In these cells induction correlates with the inhibition of CDK4 and CDK6 activities and with the induction of G1 arrest (K. DAI and D. BEACH, unpublished data). IL-6 also induces p18 expression in IgG-bearing lymphoblastoid cells and this correlates with cell cycle arrest and terminal differentiation into B-cells (MORSE et al. 1997). p18 is also highly induced during myogenic differentiation, potentially mediating accompanying growth arrest via inhibition of CDK6 and CDK4 (FRANKLIN and XIONG 1996). Considered together, these results suggest that p18 and p19 may play a role in the differentiation of some cell types. Whether these roles are critical awaits reports of p18 and p19-null mice.

p19 maps to 19p13.2, and this locus has been found as a translocation breakpoint [(1;19)(q23;p13)] in some cases of pediatric acute lymphoblastic leukemia. p18 maps to 1p32, a region that is frequently rearranged in neuroblastomas. However, neither homozygous deletion nor intragenic mutations of p18 and p19 genes have been found in cell lines or in primary tumors (LAPOINTE et al. 1996; ZARIWALA and XIONG 1996; KAWAMATA et al. 1995; OKAMOTO et al. 1995; SIEBERT et al. 1996; OTSUKI et al. 1996; SCHWALLER et al. 1997). Therefore, the INK4 family may be divided into two groups. One includes p15 and p16; in this group, genetic and epigenetic alterations might contribute to the development of human cancers. The other group includes p18 and p19 which may not play a role in tumor suppression since somatic mutation of these genes is uncommon.

References

Aprelikova O, Xiong Y, Liu ET (1995) Both p16 and p21 families of cyclin-dependent kinase (CDK) inhibitors block the phosphorylation of cyclin-dependent kinases by the CDK-activating kinase. J Biol Chem 270:18195–18197
Bar-Sagi D, Feramisco JR (1985) Microinjection of the ras oncogene protein into PC12 cells induces morphological differentiation. Cell 42:841–848
Benito M, Porras A, Nebreda AR, Santos E (1991) Differentiation of 3T3-L1 fibroblasts to adipocytes induced by transfection of ras oncogenes. Science 253:565–568
Boon LM, Mulliken JB, Vikkula M, Watkins H, Seidman J, Olsen BR, Warman ML (1994) Assignment of a locus for dominantly inherited venous malformations to chromosome 9p. Hum Mol Genet 3:1583–1587
Chan FK, Zhang J, Cheng L, Shapiro DN, Winoto A (1995) Identification of human and mouse p19, a novel CDK4 and CDK6 inhibitor with homology to p16INK4. Mol Cell Biol 15:2682–2688
Clarke AR, Maandag ER, van Roon M, van der Lugt NM, van der Valk M, Hooper ML, Berns A, te Riele H (1992) Requirement for a functional Rb-1 gene in murine development. Nature 359:328–330
Cox LS, Lane DP (1995) Tumour suppressors, kinases and clamps: how p53 regulates the cell cycle in response to DNA damage. Bioessays 17:501–508
Draetta G (1990) Cell cycle control in eukaryotes: molecular mechanisms of cdc2 activation. Trends Biochem Sci 15:378–383

Fang F, Orend G, Watanabe N, Hunter T, Ruoslahti E (1996) Dependence of cyclin E-CDK2 kinase activity on cell anchorage. Science 271:499–502

Flores JF, Walker GJ, Glendening JM, Haluska FG, Castresana JS, Rubio MP, Pastorfide GC, Boyer LA, Kao WH, Bulyk ML, Barnhill RL, Hayward NK, Housman DE, Fountain JW (1995) Loss of the p16INK4a and p15INK4b genes, as well as neighboring 9p21 markers, in sporadic melanoma. Cancer Res 56:5023–5503

Franklin DS, Xiong Y (1996) Induction of p18INK4c and its predominant association with CDK4 and CDK6 during myogenic differentiation. Mol Biol Cell 7:1587–1599

Galaktionov K, Chen X, Beach D (1996) Cdc25 cell-cycle phosphatase as a target of c-myc. Nature 382:511–517

Guan KL, Jenkins CW, Li Y, Nichols MA, Wu X, O'Keefe CL, Matera AG, Xiong Y (1994) Growth suppression by p18, a p16INK4/MTS1- and p14INK4B/MTS2-related CDK6 inhibitor, correlates with wild type pRB function. Genes Dev 8:2939–2952

Hall M, Bates S, Peters G (1995) Evidence for different modes of action of cyclin dependent kinase inhibitors: p15 and p16 bind to kinases, p21 and p27 bind to cyclins. Oncogene 11:1581–1588

Hannon GJ, Beach D (1994) p15INK4B is a potential effector of TGF-beta-induced cell cycle arrest. Nature 371:257–261

Hansen R, Oren M (1997) p53: from inductive signal to cellular effect. Curr Opin Genet Dev 7:46–51

Hara E, Smith R, Parry D, Tahara H, Stone S, Peters G (1996) Regulation of p16CDKN2 expression and its implications for cell immortalization and senescence. Mol Cell Biol 16:859–867

Hirai H, Roussel MF, Kato JY, Ashmun RA, Sherr CJ (1995) Novel INK4 proteins, p19 and p18, are specific inhibitors of the cyclin D-dependent kinases CDK4 and CDK6. Mol Cell Biol 15:2672–2681

Hunter T, Pines J (1994) Cyclin and cancer. II. Cyclin D and CDK inhibitors come of age. Cell 79:573–582

Iavarone A, Massague J (1997) Repression of the CDK activator Cdc25A and cell-cycle arrest by cytokine TGF-beta in cells lacking the CDK inhibitor p15. Nature 387:417–422

Jacks T, Fazeli A, Schmitt EM, Bronson RT, Goodell MA, Weinberg RA (1992) Effects of an Rb mutation in the mouse. Nature 359:295–300

Jen J, Harper JW, Bigner SH, Bigner DD, Papadopoulos N, Markowitz S, Wilson JK, Kinzler KW, Vogelstein B (1994) Deletion of p16 and p15 genes in brain tumors. Cancer Res 54:6353–6358.

Kalus W, Baumgartner R, Renner C, Neogel A, Chan FK, Winoto A, Holak TK (1997) NMR structural characterization of the CDK inhibitor, p19 INK4d. FEBS Lett 410:127–132

Kamb A, Gruis NA, Weaver-Feldhaus J, Liu Q, Harshman K, Tavtigian SV, Stockert E, Day RS, Johnson BE, Skolnick MH (1994) A cell cycle regulator potentially involved in genesis of many tumor types. Science 264:436–440.

Kawamata N, Miller CW, Koeffler HP (1995) Molecular analysis of a family of cyclin-dependent kinase inhibitor genes (p15/MTS2/INK4b and p18/INK4c) in non-small cell lung cancers. Mol Carcinog 14:263–268

King RW, Jackson PK, Kirschner MW (1994) Mitosis in transition. Cell 79:563–571

Koh J, Enders GH, Dynlacht BD, Harlow E (1995) Tumour-derived p16 alleles encoding proteins defective in cell-cycle inhibition. Nature 375:506–509

Lapointe J, Lachance Y, Labrie Y, Labrie C (1996) A p18 mutant defective in CDK6 binding in human breast cancer cells. Cancer Res 56:4586–4589

Lee EY, Chang CY, Hu N, Wang YC, Lai CC, Herrup K, Lee WH, Bradley A (1992) Mice deficient for Rb are nonviable and show defects in neurogenesis and haematopoiesis. Nature 359:288–294

Li JM, Nichols MA, Chandrasekharan S, Xiong Y, Wang XF (1995) Transforming growth factor beta activates the promoter of cyclin-dependent kinase inhibitor p15INK4B through an Sp1 consensus site. J Biol Chem 270:26750–26753

Lois AF, Cooper LT, Geng Y, Nobori T, Carson D (1995) Expression of the p16 and p15 cyclin-dependent kinase inhibitors in lymphocyte activation and neuronal differentiation. Cancer Res 55:4010–4013

Lu KP, Hunter T (1995) Evidence for a NIMA-like mitotic pathway in vertebrate cells. Cell 81:413–424

Lukas J, Parry D, Aagaard L, Mann DJ, Bartkova J, Strauss M, Peters G, Bartek J (1995) Retinoblastoma protein dependent cell cycle inhibition by the tumor suppressor p16. Nature 375:503–506

Massague J, Polyak K (1995) Mammalian antiproliferative signals and their targets. Curr Opin Genet Dev 5:91–96

Matsushime H, Quelle DE, Shurtleff SA, Shibuya M, Sherr CJ, Kato JY (1994) D-type cyclin-dependent kinase activity in mammalian cells. Mol Cell Biol 14:2066–2076

Medema RH, Herrera RE, Lam F, Weinberg RA (1995) Growth suppression by p16INK4 requires functional retinoblastoma protein. Proc Natl Acad Sci USA 92:6289–6293

Morse L, Chen D, Franklin D, Xiong Y, Chen-Kiang S (1997) Induction of cell cycle arrest and B cell terminal differentiation by CDK inhibitor p18(INK4c) and IL-6. Immunity 6:47–56

Okamoto A, Hussain SP, Hagiwara K, Spillare EA, Rusin MR, Demetrick DJ, Serrano M, Hannon GJ, Shiseki M, Zariwala M et al. (1995) Mutations in the p16INK4/MTS1/CDKN2, p15INK4B/MTS2, and p18 genes in primary and metastatic lung cancer. Cancer Res 55:1448–1451

Okuda T, Hirai H, Valentine VA, Shurtleff SA, Kidd VJ, Lahti JM, Sherr CJ, Downing JR (1995) Molecular cloning, expression pattern, and chromosomal localization of human CDKN2D/INK4d, an inhibitor of cyclin D-dependent kinases. Genomics 29:623–630

Otsuki T, Jaffe ES, Wellmann A, Kumar S, Condron KS, Raffeld M (1996) Absence of p18 mutations or deletions in lymphoid malignancies. Leukemia 10:356–360

Pines J (1996) Cyclin from sea urchins to HeLas: making the human cell cycle. Biochem Soc Trans 24:115–133

Polyak K, Lee MH, Erdjument-Bromage H, Koff A, Roberts JM, Tempst P, Massague J (1994) Cloning of p27Kip1, a cyclin-dependent kinase inhibitor and a potential mediator of extracellular antimitogenic signals. Cell 78:59–66

Quelle DE, Ashmun RA, Hannon GJ, Rehberger PA, Trono D, Richter KH, Walker C, Beach D, Sherr CJ, Serrano M (1995a) Cloning and characterization of the murine p16 INK4a and p15 INK4b genes. Oncogene 11:635–645

Quelle DE, Zindy F, Ashmun RA, Sherr CJ (1995b) Alternative reading frames of the INK4a tumor suppressor gene encode two unrelated proteins capable of inducing cell cycle arrest. Cell 83:993–1000

Reynisdottir I, Massague J (1997) The subcellular locations of p15(Ink4b) and p27(Kip1) coordinate their inhibitory interactions with cdk4 and cdk2. Genes Dev 11:492–503

Reynisdottir I, Polyak K, Iavarone A, Massague J (1995) Kip/Cip and Ink4 Cdk inhibitors cooperate to induce cell cycle arrest in response to TGF-beta. Genes Dev 9:1831–1845

Sandhu C, Garbe J, Bhattacharya N, Daksis J, Pan CH, Yaswen P, Koh J, Slingerland JM, Stampfer MR (1997) Transforming growth factor beta stabilizes p15INK4B protein, increases p15INK4B-cdk4 complexes, and inhibits cyclin D1-cdk4 association in human mammary epithelial cells. Mol Cell Biol 17:2458–2467

Sangfelt O, Erickson S, Einhorn S, Grander D (1997) Induction of Cip/Kip and Ink4 cyclin dependent kinase inhibitors by interferon-alpha in hematopoietic cell lines. Oncogene 14(4):415–423

Schwaller J, Pabst T, Koeffler HP, Niklaus G, Loetscher P, Fey MF, Tobler A (1997) Expression and regulation of G1 cell-cycle inhibitors (p16INK4A, p15INK4B, p18INK4C, p19INK4D) in human acute myeloid leukemia and normal myeloid cells. Leukemia 11:54–63

Serrano M, Hannon GJ, Beach D (1993) A new regulatory motif in cell-cycle control causing specific inhibition of cyclin D/CDK4. Nature 366 (6456):704–707

Serrano M, Gomez-lahoz E, DePinho R, Beach D, Bar-Sagi D (1995) Inhibition of ras-induced proliferation and cellular transformation by p16. Science 267:249–252

Serrano M, Lee H, Chin L, Cordon-Cardo C, Beach D, DePinho RA (1996) Role of the INK4a locus in tumor suppression and cell mortality. Cell 85:27–37

Serrano M, Lin AW, McCurrach ME, Beach D, Lowe SW (1997) Oncogenic ras provokes premature cell senescence associated with accumulation of p53 and p16INK4a. Cell 88:593–602

Sherr CJ (1993) Mammalian G1 cyclins. Cell 73:1059–1065

Sherr CJ (1994) G1 phase progression: cycling on cue. Cell 79:551–555

Sherr CJ, Roberts JM (1995) Inhibitors of mammalian G1 cyclin-dependent kinases. Genes Dev 9: 1149–1163

Siebert R, Willers CP, Opalka B (1996) Role of the cyclin-dependent kinase 4 and 6 inhibitor gene family p15, p16, p18 and p19 in leukemia and lymphoma. Leuk Lymphoma 23:505–520

Skapek SX, Rhee J, Spicer DB, Lassar AB (1995) Inhibition of myogenic differentiation in proliferating myoblasts by cyclin D1-dependent kinase. Science 267(5200):1022–1024

Tam SW, Theodoras AM, Shay JW, Daetta GF, Pagano M (1994) Differential expression and regulation of cyclin D1 protein in normal and tumor human cells: association with CDK4 is required for cyclin D1 function in G1 progression. Oncogene 9:2663–2674

Wang J, Walsh K (1996) Resistance to apoptosis conferred by Cdk inhibitors during myocyte differentiation. Science 273:359–361

Weinberg RA (1995) The retinoblastoma protein and cell cycle control. Cell 81:323–330

Wolfel T, Hauer M, Schneider J, Serrano M, Wolfel C, Klehmann-Hieb E, De Plaen E, Hankeln T, Meyer zum Buschenfelde KH, Beach D (1995) A p16INK4a-insensitive CDK4 mutant targeted by cytolytic T lymphocytes in a human melanoma. Science 269:1281–1284

Zariwala M, Xiong Y (1996) Lack of mutation in the cyclin-dependent kinase inhibitor, p19INK4d, in tumor-derived cell lines and primary tumors. Oncogene 13:2033–2038

Zariwala M, Liu E, Xiong Y (1996) Mutational analysis of the p16 family cyclin-dependent kinase inhibitors p15INK4b and p18INK4c in tumor-derived cell lines and primary tumors. Oncogene 12:451–455

Zhang S, Endo S, Koga H, Ichikawa T, Feng X, Onda K, Washiyama K, Kumanishi T (1996) A comparative study of glioma cell lines for p16, p15, p53 and p21 gene alterations. Jpn J Cancer Res 87:900–907

Zhou JN, Linder S (1996) Expression of CDK inhibitor genes in immortalized and carcinoma derived breast cell lines. Anticancer Res 16:1931–1935

Zhu X, Ohtsubo M, Bohmer RM, Roberts JM, Assoian RK (1996) Adhesion-dependent cell cycle progression linked to the expression of cyclin D1, activation of cyclin E-cdk2, and phosphorylation of the retinoblastoma protein. J Cell Biol 133(2):391–403

Role of Cyclin-Dependent Kinases and Their Inhibitors in Cellular Differentiation and Development

S.P. Chellappan[1], A. Giordano[3], and P.B. Fisher[1,2]

1	Introduction	58
2	Cyclin-Dependent Kinase Inhibitor p21	63
2.1	Identification and Cloning	63
2.2	Cellular Differentiation	64
2.2.1	Human Melanoma Differentiation	64
2.2.2	Hematopoietic Cell Differentiation	68
2.2.3	Muscle Cell Differentiation	71
2.2.4	Keratinocyte Differentiation	72
2.2.5	Development	73
2.2.6	Mode of Action	74
3	Cyclin-Dependent Kinase Inhibitor p27^{KIP1}	76
3.1	Regulation of p27 Levels in Cells	79
3.2	Role in Proliferation and Differentiation	79
3.3	Role in Development	82
3.3.1	General Phenotype of p27-Null Mice	83
3.3.2	Abnormalities of the Thymus in p27-Null Mice	83
3.3.3	Pituitary Hyperplasia in p27$^{-/-}$	84
3.3.4	Reproductive System Abnormalities in p27$^{-/-}$ Mice	85
3.3.5	Effects of p27 Knockout on the Retinal Structure	85
3.3.6	Growth Arrest Is Normal in p27$^{-/-}$ Fibroblasts	86
3.3.7	Mechanism of Cyclin-Dependent Kinase Inhibition by p27	86
4	Cyclin-Dependent Kinase Inhibitor p57KIP2	86
4.1	Differentiation and Development	87
5	INK4 Family of Cyclin-Dependent Kinase Inhibitors	88
5.1	p16 and Cellular Differentiation	90
5.2	Role of p16 in Development	90
5.3	Tumor Susceptibility of p16-Null Mice	91
5.4	Properties of p16$^{-/-}$ Mouse Embryo Fibroblasts	91
6	p18INK4C	92
7	p19INK4D	92
8	Future Outlook	93
References		94

[1]Department of Pathology, Herbert Irving Comprehensive Cancer Center, College of Physicians and Surgeons of Columbia University, 630 West 168th St, New York, NY 10032, USA

[2]Department of Urology, Herbert Irving Comprehensive Cancer Center, College of Physicians and Surgeons of Columbia University, New York, NY 10032, USA

[3]Department of Pathology, Anatomy and Cell Biology, Kimmel Cancer Institute, Thomas Jefferson University, Sbarro Institute for Cancer Research and Molecular Medicine, 233 South 10th Street, Philadelphia, PA 19107, USA

1 Introduction

The proliferation of normal cells is regulated by a combination of stimulatory and inhibitory factors that can respond to external signals in a coordinated manner (MacLachlan et al. 1995; Grana and Reddy 1995; Paggi et al. 1996). Various mitogens, growth factors, cytokines, and a host of other agents can perturb the cell cycle machinery eliciting proliferation, differentiation, or apoptosis (Reed et al. 1994; Reddy 1994). Any permanent alteration of the cell cycle-regulatory mechanisms can lead to abnormal proliferation, resulting in neoplasia (Bishop 1987, 1991; Morgan 1992). Such a situation can result following the inactivation of various tumor suppressor proteins, such as Rb and its family members and p53 (Marshall 1991; Cobrinik et al. 1992; Hollingsworth et al. 1993; Ewen 1994; Hooper 1994; Picksley and Lane 1994; Weinberg 1995; Giordano and Kaiser 1996), or by the dominant activation of positive-acting components of the cell cycle machinery, such as the cyclins and cyclin-dependent kinases (CDK) or proto-oncogenes (Pines 1995a, b; Kamb 1995; Bates and Peters 1995).

Normal quiescent cells are maintained in the G_0 phase of the cell cycle and can progress through the cell cycle upon receiving the appropriate mitogenic stimuli (Fig. 1; for a review, see Sherr 1994). These cells are constrained from proliferating in the absence of such stimuli by the tumor suppressor proteins, which prevent the transition from G_1 to S phase of the cell cycle (Sang et al. 1995; Whyte 1995; Massague and Polyak 1995). Once the cells have passed the G_1/S-phase boundary, they can complete the cell cycle since there are no other major barriers preventing this transition (Reed et al. 1994; Sherr 1994). Studies over the past few years have provided new insights on how the positive-acting components of the cell cycle, the cyclins and the CDK, can inactivate the tumor suppressor proteins and overcome the cell cycle block (Pines 1995b; Paggi et al. 1996).

The major cyclins that function in G_1 phase of the cell cycle are cyclins D and E, although cyclin A and associated activity appear at the end of G_1 phase (Table 1, Fig. 1). Since their cloning and characterization just a few years ago, the D-type cyclins have attracted considerable attention, since they were found to be the first cyclins to function upon mitogen stimulation (for reviews, see Sherr 1993, 1995; Matsushime et al. 1994). The importance of D-type cyclins and their associated kinases, i.e., cdk4 and cdk6, was confirmed when it was found that it was mainly through their activity that phosphorylation and inactivation of the Rb protein occurs (Ewen et al. 1993; Kato et al. 1993). Even though cyclin E–cdk2 activity can phosphorylate Rb subsequent to the phosphorylation by D-type cyclins and cdk4/cdk6, it is fairly well established that the major inactivating phosphorylation of Rb is by the latter molecule (see Weinberg 1995). Further, it has become evident that the only known substrate of D-type cyclins and associated kinases is the Rb protein, and the phosphorylation and inactivation of Rb by cdk4/cdk6 are essential for the cells to progress through the cell cycle (Ewen et al. 1993; Kato et al. 1993).

The CDK activities are mandatory for cell cycle progression and are probably the most important positive-acting molecules involved in its regulation (Fig. 1; for a

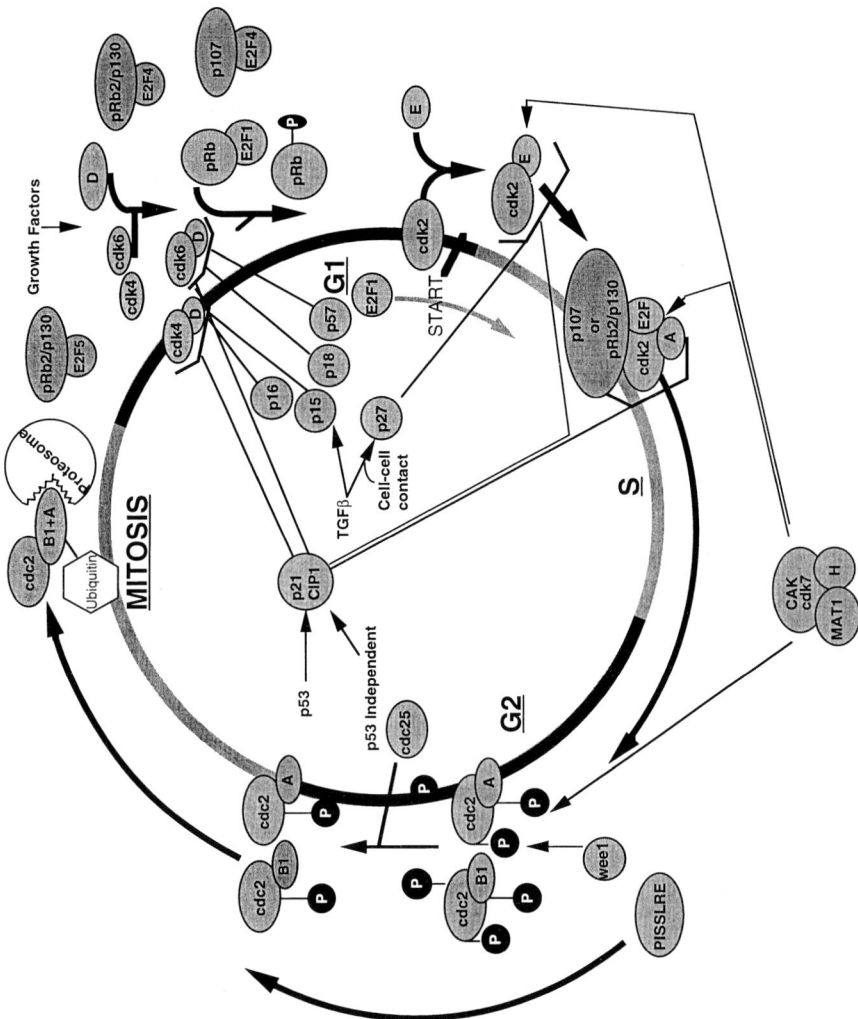

Fig. 1. Proteins involved in regulating and controlling cell cycle, including recent data indicating a possible role of PISSLRE in G_2 or G_2/M phases of the cell cycle, as a dominant negative mutant form of PISSLRE significantly increases the number of cells in G_2 phase in a manner similar to the cdc2 kinase *TGF*, transforming growth factor; *CAK*, cyclin-dependent kinase-activating kinase. (From LI et al. 1995)

review, see MORGAN 1995). Given their significant role in cell cycle progression, the activity of CDK must be regulated very precisely by multiple mechanisms. First, they have to associate with cyclins for kinase activity (PINES 1995b; POON and HUNTER 1995; MORGAN 1995). Since the levels of CDK remain relatively constant through out the cell cycle, it is the relative abundance of the cyclins that to a great extent determines the amount of CDK activity. Studies have shown that association with cyclins is not the only major positive determinant of CDK activity, but that

Table 1. Cell cycle proteins involved in cell cycle progression

Cell cycle protein	Cell cycle involvement	Function of protein	Regulation of activity	Possible involvement in cancer
Cyclin A	Through S progression into M	Activate cdk2 Activate cdc2	Transcriptional and ubiquitin degradation	Hepatocellular carcinoma and Associated with E1A
Cyclin B_1	Progression into M	Activate cdc2	Transcriptional and ubiquitin degradation	?
Cyclin B_2	Progression into M	?	Ubiquitin degradation	?
Cyclin C	Through G_1?	Activate cdk8	?	?
Cyclins D_1, D_2, D_3	Through G_1	Activate cdk4 and cdk6	Transcriptional and PEST-targeted proteolysis	D1 is PRAD1/BCL1 proto-oncogene Rearrangements and overexpression in breast carcinomas
Cyclin E	Progression into S	Activate cdk2	Transcriptional and PEST-targeted proteolysis	?
Cyclin F	Through G_2 or progression into M	Binds Skp1, involved in proteolysis	Transcriptionally activated?	Implicated in p53 response pathways
Cyclin G	Unknown	p53-responsive protein in DNA damage	Transcriptional	?
Cyclin H	Ubiquitous	Activates cdk7	?	?
Cyclin I	Mostly in G_0	Preserve G_0 arrest in terminal differentiation?	?	?
Cyclin J	Unknown	Early Drosophila embryogenesis	Transcriptional	?
cdc2	Progression into M	Phosphorylates histone H_1, Lamins, etc.	Activated by A and B cyclins	?
cdk2	Progression past START through S	Associates with and phosphorylates DNA replication machinery	Activated by A and E cyclins	?
cdk3	Through G_1 or progression past START	?	?	Located near LOH of BRCA1
cdk4	Through G_1	Phosphorylates pRb tumor suppressor	Activated by D_1, D_2, and D_3 cyclins	Gene amplification in gliomas

Role of Cyclin-Dependent Kinases and Their Inhibitors

cdk5	In G_0?	Phosphorylates *tau* and neurofilament brain proteins	Activated by p35 subunit	?
cdk6	Through G_1	Phosphorylates pRb tumor suppressor	Activated by D_1, D_2 and D_3 cyclins	Non-Hodgkin's lymohoma?
cdk7	Ubiquitous	Activate cdk and phosphorylates CTD of RNA polymerase II	Activity dependent on presence of substrate?	?
cdk8	Ubiquitous	Phosphorylate CTD of RNA polymerase II	Binding of cyclin C	?
PITSLRE	G_1?	Apoptosis	?	Localized in region deleted in several carcinomas
PISSLRE	G_2?	?	?	Localized near LOH in breast and prostate carcinomas
PITALRE	G_1?	Phosphorylates pRb?	?	Non-Hodgkin's lymphoma?
p21$^{WAF1/CIP1/MDA6}$	Mostly in G_1 and S	Inhibits cdk2, cdk4, and cdk6 PCNA and DNA replication	Transcribed by tumor suppressor p53	Associated with p53 loss in cancer
p16^{INK4A}	G_1	Inhibits cdk4 and cdk6 by cyclin D displacement	Transcribed by pRb-associated transcription factor	Multiple tumor suppressor 1 locus
p27^{Kip1}	G_1	Inhibits cdk2 upon TGF-β or cell–cell contact	Diluted distribution among cdk until relocalized	Localized in regions involved in leukemia
p15^{INK4B}	G_1	Inhibits cdk4 and cdk6 upon TGF-β treatment	Transcribed downstream of TGF-β exposure	Multiple tumor suppressor 2 locus
p18^{INK4C}	G_1	Inhibits cdk4 and cdk6	?	?
p19^{INK4D}	G_1	Inhibits cdk4 and cdk6	?	?

PEST, proline, glutamic acid, serine and threonine; CTD, carboxyl terminal domain; PCNA, proliferating cell nuclear antigen; TGF, transforming growth factor; LOH, loss of heterozygosity (Li et al. 1995).

specific phosphorylation on a conserved threonine residue (T160) is also a major determinant of CDK function (Y. GU et al. 1992; MORGAN and DEBONDT 1994; DESAI et al. 1995). Surprisingly, it has been established that the CDK-activating kinase (CAK) that is involved in the phosphorylation of CDK is itself a distinct cyclin-CDK complex, cyclin H–cdk7 (POON et al. 1993; KATO et al. 1994b; M. MATSUOKA et al. 1994; FISHER and MORGAN 1994). The CAK is apparently a multifunctional kinase; it has been implicated in the excision–repair pathway of DNA and has been shown to be a component of the basal transcription factor TFIIH (SHEIKHATTAR et al. 1995). Its function in the repair pathway is not clear, but its presence in the TFIIH complex has led to the speculation that it is the major kinase involved in the phosphorylation of the C terminus of the large subunit of RNA polymerase II, a step that is required for promoter clearance and elongation.

In addition to activation by phosphorylation on Thr-160, the CDK must be dephosphorylated at other residues to elicit appropriate function. Studies on yeast and on mammalian cdk2 have shown that dephosphorylation of Thr-14 and Tyr-15 is required for full CDK activity (for a review, see MORGAN 1995). These two residues reside in the ATP-binding domain of the CDK, and removal of the phosphate appears to be necessary for proper kinase activity. The phosphatases involved in this activation process have not been identified, but it has been postulated that in mammals the cdc25 protein might be involved. This fits very well with the kinetics of cdc25 activation, which precedes the activation of CDK (GAUTIER et al. 1991; JINNO et al. 1994; GALAKTIONOV et al. 1995).

A new class of proteins identified and cloned in 1993 turned out to be the first known negative regulators of CDK activity. Such proteins, termed CDK inhibitors (CDI), have been shown to be as important as cyclins in regulating CDK function and hence appear to play a major role in cell cycle regulation (for reviews, see SHERR and ROBERTS 1995; MACLACHLAN et al. 1995). Because of their function at a key regulatory point of the cell cycle, CDI have been implicated in the regulation of many cellular processes, such as differentiation, senescence, and apoptosis (JIANG and FISHER 1993; NODA et al. 1994; MISSERO and DOTTO 1996; WALSH et al. 1994; WANG and WALSH 1996; JIANG et al. 1996a; ROBETORYE et al. 1996; LIU et al. 1996). All the CDI cloned to date function by inducing growth arrest at the G_1/S-phase boundary very much like the known tumor suppressor genes (Fig. 1). Indeed, inactivation of many of the CDI have been detected in human tumors, suggesting that they can function as true tumor suppressors (BIGGS and KRAFT 1995; KAMB 1995).

Two major classes of CDI have been identified. The first consists of p21/Waf1/Cip1/Sdi1/*mda*-6, p27Kip1, and p57Kip2 (Table 1; for a review, see SHERR and ROBERTS 1995), which can act on most of the cyclin–CDK complexes and even on kinases unrelated to CDK. These molecules interact with cyclin–CDK complexes, leading to an inactivation of the kinase activity. The second family of CDI consists of the INK4 family of inhibitors, i.e., p16INK4a, p15INK4b, p18INK4c, and p19INK4d. The naming of these genes reflects the fact that they preferentially inhibit CDK4 kinase activity. Their mode of action is different from the p21 family in that they bind to the CDK and prevent binding to the cyclins (HALL et al. 1995).

As anticipated, the CDK play a major role in the processes of cell cycle regulation, differentiation, and development, and this review will attempt to provide an overview of the literature in this rapidly expanding field. The relative role of each CDI in differentiation and development will also be reviewed.

2 Cyclin-Dependent Kinase Inhibitor p21

2.1 Identification and Cloning

The orderly transit of cells through the cell cycle is critical for the maintenance of homeostasis in both prokaryotic and eukaryotic organisms (for a review, see SHERR 1993; HARTWELL and KASTAN 1994; HUNTER and PINES 1994; ELLEDGE and HARPER 1994; SANG et al. 1995; PINES 1995c; CLURMAN and BRUNS 1995; BIGGS and KRAFT 1995; MACLACHLAN et al. 1995; SHERR and ROBERTS 1995). In specific contexts, such as terminal cell differentiation and cellular senescence, a permanent exit from the cell cycle is mandatory (GOLDSTEIN 1990; NODA et al. 1994; JIANG et al. 1994a, 1996a; ROBETORYE et al. 1996; SMITH and PEREIRA-SMITH 1996). Fidelity of cell cycle control contributes directly to preservation of a normal cellular phenotype, whereas abnormalities in this process link cell cycle and cancer (HARTWELL and KASTAN 1994; JIANG et al. 1994a; SANG et al. 1995; CLURMAN and BRUNS 1995; BIGGS and KRAFT 1995; MACLACHLAN et al. 1995).

Cell cycle progression results from the coordinated expression and interaction of positive-regulatory proteins, including cyclins and their associated CDK, and negative-regulating proteins, CDI (for a review, see SHERR 1993; HUNTER and PINES 1994; BIGGS and KRAFT 1995; SHERR and ROBERTS 1995). The CDI first identified and most extensively studied is p21, also referred to as CAP20, Cip1, PIC1, *mda*-6, SDI1, and WAF1 (XIONG et al. 1993a; HARPER et al. 1993; EL-DEIRY et al. 1993; W. GU et al. 1993; JIANG and FISHER 1993; NODA et al. 1994; JIANG et al. 1994a, 1995b, 1996a). The importance of p21 in a wide range of cellular activities is reflected by the diversity of approaches that led to the identification and cloning of this gene. HARPER et al. (1993) detected and cloned p21 as a CDK2-interacting protein, Cip1, using the yeast two-hybrid system. Biochemical purification approaches resulted in the cloning of p21 (XIONG et al. 1993a) and CAP20 (W. GU et al. 1993) as a protein interacting with CDK2. JIANG and FISHER (1993) identified and cloned p21 as a melanoma differentiation-associated gene, *mda*-6, that is up-regulated as a consequence of growth suppression and induction of terminal cell differentiation in human melanoma cells. EL-DEIRY et al. (1993) identified p21 using human glioblastoma cells containing an inducible wild-type p53 tumor suppressor gene and subtraction hybridization, referred to as wild-type p53-activated fragment 1 (WAF1). An expression screening method to identify and isolate cDNA coding for inhibitors of DNA synthesis from senescent human diploid fibroblasts resulted in the cloning of p21 as senescent-derived inhibitor-1, SDI1 (NODA et al.

1994). A number of recent studies support an involvement of p21 in numerous important cellular processes, including DNA replication, mitogenesis, DNA repair, differentiation, and development (for a review, see in HUNTER and PINES 1994; ELLEDGE and HARPER 1994; SANG et al. 1995; PINES 1995c; MACLACHLAN et al. 1995; SHERR and ROBERTS 1995; XIONG et al. 1992).

2.2 Cellular Differentiation

2.2.1 Human Melanoma Differentiation

The first demonstration of an association between cellular differentiation and p21 came from studies using human melanoma cells (JIANG and FISHER 1993; JIANG et al. 1994a). When exposed to the combination of recombinant human fibroblast interferon (IFN-β) and the antileukemic compound mezerein (MEZ), human melanoma cells lose proliferative ability and terminally differentiate (P.B. FISHER et al. 1985; JIANG et al. 1993). To elucidate the mechanism by which IFN-β + MEZ alters cell growth and induce terminal differentiation in human melanoma cells, JIANG and FISHER (1993) used a sensitive and efficient subtraction hybridization procedure. cDNA libraries were constructed from H0-1 cells treated with IFN-β + MEZ for 2, 4, 8, 12, and 24 h (Ind^+; tester library) and from logarithmically growing H0-1 cells (Ind^-; driver library). The Ind^+ and Ind^- cDNA libraries were directionally cloned into the λ Uni-ZAP phage vector. Subtraction hybridization was then conducted between double-stranded tester DNA and single-stranded driver DNA prepared by mass excision of the libraries. The subtracted cDNA were subsequently cloned into λ Uni-ZAP phage vector and used to screen northern blots containing total RNA isolated from untreated H0-1 cells (control) or H0-1 cells treated for 24 h with IFN-β, MEZ, or IFN-β + MEZ. This approach led to the identification of *mda* cDNA displaying elevated expression as a function of treatment with the different inducing agents (JIANG and FISHER 1993). The *mda*-6 cDNA is the CDI p21 (JIANG et al. 1994a, 1995b).

Treatment of human melanoma cells with IFN-β + MEZ for 24 h results in the induction or enhanced expression of *mda*-6 (p21) RNA and protein (JIANG et al. 1995b). In terminally differentiated melanoma cells, the levels of *mda*-6 (p21) remain elevated, suggesting that this gene product may be a component of the growth arrest and terminal differentiation process. Induction of *mda*-6 (p21) in human melanoma cells is not restricted to differentiation, since growth to high saturation densities, treatment with DNA-damaging agents (such as methyl methanesulfonate), or incubation in medium lacking serum elevates *mda*-6 (p21) RNA levels (JIANG et al. 1995b). The process of augmented *mda*-6 (p21) expression is rapid and occurs in the absence of protein synthesis (presence of cycloheximide), suggesting that p21 induction is an immediate-early response to agents inducing growth arrest, DNA damage, and differentiation.

Although initially considered to be dependent on wild-type p53 for induction (EL-DEIRY et al. 1993), it is now apparent that p21 induction can occur by both p53-dependent and p53-independent mechanisms (JIANG et al. 1994b; STEINMAN et al. 1994; MICHIELI et al. 1994; MACJOHNSON et al. 1994; ELBENDARY et al. 1994; SHEIKH et al. 1994; PARKER et al. 1995; ZHANG et al. 1995; DIGIUSEPPE et al. 1995; MACLEOD et al. 1995; DENG et al. 1995). Although mutations in p53 are the most common genetic alteration in most human cancers (HOLLENSTEIN et al. 1991), this does not apply to human melanoma (VOLKENANDT et al. 1991; CASTRESANA et al. 1993; GREENBLATT et al. 1994; MONTANO et al. 1994; JIANG et al. 1995b). In fact, evidence is now accumulating that many human melanomas contain immunologically wild-type p53 (VOLKENANDT et al. 1991; CASTRESANA et al. 1993; LOGANZO et al. 1994; GREENBLATT et al. 1994; MONTANO et al. 1994; JIANG et al. 1995b). These findings represent an interesting paradox, suggesting that melanomas may be unique among cancers since they can coexist and evolve to more aggressive stages even in the presence of elevated levels of nuclear-localized wild-type p53 protein. Alternatively, the wild-type p53 protein in metastatic melanoma may be functionally inactivated by interacting with cellular proteins. It is also possible that the wild-type p53 in metastatic melanoma may not be able to transcriptionally modify specific target genes necessary for inhibiting cancer progression or that the downstream wild-type p53 target genes themselves are defective in metastatic human melanoma (LOGANZO et al. 1994; JIANG et al. 1995b).

Experiments have evaluated p53 and p21 (*mda*-6) expression during the processes of growth arrest and terminal differentiation in human melanoma cells (JIANG et al. 1995b, c). When H0-1 human melanoma cells are induced to reversibly differentiate (MEZ treatment) or terminally differentiate (IFN-β + MEZ treatment), p53 mRNA and protein levels decrease while *mda*-6 (p21) mRNA and protein levels increase (JIANG et al. 1995b, c). These experiments demonstrate an inverse relationship between p53 and p21 (*mda*-6) expression in the process of growth arrest and terminal differentiation in H0-1 cells. The high levels of presumably wild-type p53 in human melanoma cells with the corresponding low levels of p21 (*mda*-6) clearly indicate a lack of dependence on wild-type p53 expression for p21 induction in human melanoma cells. This may occur because wild-type p53 induces downstream genes that may actually function as inhibitors of p21 (*mda*-6) expression in melanoma cells. Alternatively, the wild-type p53 may be inactive, or downstream pathways responsive to wild-type p53 may be defective in human melanoma cells. It is also possible that melanocytes, nevi, and early-stage primary melanomas contain reduced levels of a functional wild-type p53 that mediates p21 (*mda*-6) expression, whereas the presence of predominantly inactive wild-type p53 in melanomas prevents p21 (*mda*-6) expression. When metastatic melanoma cells differentiate after exposure to MEZ or IFN-β + MEZ, the predominantly inactive wild-type p53 may be reduced, thereby paradoxically resulting in restoration of wild-type p53 function, elevated p21 levels, and growth arrest.

Studies employing human melanoma indicate that *mda*-6 (p21) may function as a negative regulator of melanoma progression (JIANG et al. 1995b, 1996a). Malignant melanoma, with the exception of nodular-type melanoma, is a progressive disease characterized by temporally defined stages, including melanocyte, nevus, dysplastic nevus, radial growth-phase (RGP) primary melanoma, vertical growth-phase (VGP) primary melanoma, and metastatic melanoma (CLARK 1991; KERBEL 1990, HERLYN 1990; LU and KERBEL 1994; JIANG et al. 1994a). The mechanism controlling melanoma progression is not currently known. One scheme proposed to explain this process, the *aberrant differentiation/modified gene expression model*, suggests that metastatic melanoma cells develop as a consequence of anomalous patterns of gene expression resulting in abnormal differentiation (JIANG et al. 1994a, 1996a). Metastatic melanomas fail to express specific genes normally involved in controlling cell proliferation and expression of differentiated properties (JIANG et al. 1994a, 1996a). In contrast, the appropriate growth- and differentiation-regulating genes are expressed in melanocytes, nevi, RGP primary melanomas, and/or VGP primary melanomas (JIANG et al. 1994a, 1996a). When exposed to IFN-β + MEZ, irreversible growth arrest and terminal differentiation result in metastatic melanoma, presumably because the appropriate genes negatively regulating growth and the cancer phenotype are activated.

If the aberrant differentiation/modified gene expression hypothesis is correct, human melanocytes and terminally differentiated human melanoma cells should express specific genes not expressed or expressed at reduced levels in actively proliferating metastatic melanomas. This possibility is supported by the cloning of *mda*-6 (p21), and the novel gene *mda*-7, by subtraction hybridization from a differentiation inducer (IFN-β + MEZ)-treated H0-1 human melanoma library (JIANG and FISHER 1993; JIANG et al. 1994a, 1995a, b, 1996a). Several lines of evidence suggest that *mda*-6 (p21) and *mda*-7 may function as negative regulators of growth and progression in human melanomas. Treatment of metastatic human melanoma cells with IFN-β + MEZ results in a loss of proliferative potential and terminal differentiation in human melanoma cells (P.B. FISHER et al. 1985; JIANG et al. 1993). This process correlates with a rapid and persistent elevation in the levels of both *mda*-6 (p21) and *mda*-7 mRNA and protein (JIANG et al. 1995a, b). In contrast, the levels of *mda*-6 (p21) and *mda*-7 mRNA are higher in melanocytes than in actively proliferating primary and metastatic human melanomas (JIANG et al. 1995a, b, 1996a). The higher levels of these *mda* genes in melanocytes may mediate the slower growth rate exhibited by these cells versus melanoma cells. In this context, loss of regulation of *mda*-6 (p21) and/or *mda*-7 may directly mediate enhanced growth potential, a loss of differentiated properties, and cancer progression.

Recent studies suggest that it may be possible to spontaneously progress early-stage, nontumorigenic, or poorly tumorigenic (in nude mice) primary human melanoma cells to a more progressed tumorigenic phenotype by injecting cells into nude mice in combination with Matrigel (basement membrane matrix) (MACDOUGALL et al. 1993; KOBAYASHI et al. 1994). Using this model system, the levels of *mda*-6 (p21) and *mda*-7 decrease as a function of progression. A direct test of the effect of *mda*-6 (p21) and *mda*-7 on progression will mandate a functional analysis of the effect of either elevating or inhibiting expression of these *mda* genes on cellular phenotype. When ectopically expressed in human melanoma cells, *mda*-6 (p21) and *mda*-7 suppress growth and the transformed phenotype (JIANG et al. 1995a, b, 1996b; YANG et al. 1995). Using an adenoviral vector expressing p21,

YANG et al. (1995) demonstrated the induction of a more differentiated phenotype, an accumulation of cells in G_0/G_1 phase, a suppression in malignant growth in vivo, and an inhibition in the growth of preexisting tumors. These studies do not demonstrate that the mechanism involved in melanoma growth suppression is directly mediated by induction of terminal differentiation. Moreover, when an *mda*-6 (p21) expression construct is transfected into H0-1 human melanoma cells, growth is suppressed and differentiation is augmented, but the majority of these cells do not terminally differentiate. It is possible that the expression of multiple *mda* genes is necessary to induce terminal differentiation in human melanoma cells. This hypothesis is currently being tested.

The final stage of cancer progression is acquisition of metastatic potential by the evolving tumor cells (LIOTTA et al. 1991; SU et al. 1993, 1995). In progressing or late-stage cutaneous malignant melanoma, structural alterations in chromosome 6 are a common occurrence (PATHAK et al. 1983; FOUNTAIN et al. 1990; KACKER et al. 1990; TRENT 1991; DRACOPOLI et al. 1992; X. GUAN et al. 1992). Chromosome transfer studies in human melanoma provide direct functional evidence for tumor and metastasis suppressor genes on chromosome 6 (TRENT et al. 1990; WELCH et al. 1994; JIANG et al. 1995b; MIELE et al. 1996; WELCH and RIEBER 1996). When a normal chromosome 6 is transferred into the human melanoma cell lines UACC-903 and UACC-091, tumorigenicity is suppressed (TRENT et al. 1990). In contrast, insertion of a normal chromosome 6 into the tumorigenic and metastatic C8161 human cell line (WELCH et al. 1991) results in an extinction of the metastatic phenotype with retention of tumorigenic potential (WELCH et al. 1994; JIANG et al. 1995b; MIELE et al. 1996). Similarly, transfer of chromosome 6 into the tumorigenic and metastatic MelJuso human melanoma cell line results in suppression of only the metastatic properties of these tumor cells (MIELE et al. 1996). These observations support the idea that chromosome 6 contains a suppressor gene that is capable of suppressing metastasis and reverting C8161 and MelJuSo cells to a less progressed stage in melanoma evolution. If *mda*-6 (p21) expression is associated with the progression phenotype, it would be predicted that the level of this *mda* gene would increase in C8161 and MelJuSo cells containing chromosome 6. This possibility has been directly tested and confirms an increase in *mda*-6 (p21) expression in tumorigenic, but non-metastatic C8161 and MelJuSo cells containing chromosome 6 (JIANG et al. 1995b; MIELE et al. 1996). Moreover, when treated with IFN-β + MEZ, growth is suppressed and the levels of *mda*-6 (p21) increase in both C8161 and in C8161 clones containing chromosome 6 (JIANG et al. 1995b). These studies provide compelling evidence for the presence of melanoma metastasis-suppressing elements on chromosome 6 (WELCH et al. 1994; JIANG et al. 1995b; WELCH and RIEBER 1996; MIELE et al. 1996). Previous studies indicate that p21 is encoded on chromosome 6p21.2 (EL-DEIRY et al. 1993). It is possible that *mda*-6 (p21), by itself or in combination with additional genes present on chromosome 6, prevents metastatic progression in human melanoma. Alternatively, p21 (*mda*-6) may not be the critical metastasis suppressor gene on chromosome 6, and the changes in p21 (*mda*-6) expression may be associated with, rather than being the cause of, metastasis suppression. Further functional studies are necessary to definitively prove that p21 (*mda*-6) is the genetic element on chromosome 6 that is defective in human metastatic melanoma and that controls the metastatic properties of human melanoma cells.

2.2.2 Hematopoietic Cell Differentiation

The HL-60 human promyelocytic leukemia cell line can be induced to differentiate along the granulocyte or the macrophage/monocyte pathway by treatment with appropriate agents (COLLINS 1987). Exposure to phorbol ester tumor promoters, such as 12-*0*-tetradecanoyl phorbol-13-acetate (TPA) or 1,25-dihydroxyvitamin D_3 (Vit D_3), induces macrophage/monocyte differentiation (HUBERMAN and CALLAHAM 1979; LOTEM and SACHS 1979; ROVERA et al. 1979; MIYAURA et al. 1981; MCCARTHY et al. 1983; MANGELSDORF et al. 1984). In contrast, exposure to retinoic acid (RA) or dimethylsulfoxide (DMSO) induces HL-60 cells to differentiate into granulocytes (COLLINS et al. 1978; BREITMAN et al. 1980). This cell line lacks endogenous p53 genes (WOLF and ROTTER 1985). In these contexts, the HL-60 cell culture system represents a useful model for defining gene expression changes associated with macrophage/monocyte or granulocyte differentiation that occur independently of wild-type p53 expression.

After cloning *mda*-6 from a human melanoma differentiation inducer-treated subtracted cDNA library, the expression pattern of this gene was determined in additional differentiating cell culture model systems. This analysis indicates that *mda*-6, subsequently shown to be p21 (WAF1/CIP1), is an immediate-early response gene induced during differentiation of HL-60 cells along granulocyte or macrophage/monocyte pathways (JIANG et al. 1994b). Treatment of HL-60 cells with TPA and Vit D_3, which induce macrophage/monocyte differentiation, or RA and DMSO, which induce granulocyte differentiation, induces p21 mRNA expression within 1–3 h (JIANG et al. 1994b). This early induction of p21 is not apparent in HL-60 variants resistant to growth suppression and differentiation induction (JIANG et al. 1994b). Immunoprecipitation analysis with p21-specific antibody confirms that the increase in mRNA expression in HL-60 cells resulting from TPA, DMSO, or RA treatment corresponds with a temporal increase in p21 protein levels (JIANG et al. 1994b). STEINMAN et al. (1994) also demonstrated a rapid increase in p21 mRNA in HL-60 cells treated with butyrate, TPA, DMSO, and RA. In both studies, induction of p21 in HL-60 cells was shown not to depend on continuing protein synthesis, since it occurs in the presence of cycloheximide. These results document that p21 induction can occur independent of p53 and that this response is an immediate-early response to induction of both macrophage/monocyte and granulocyte differentiation in HL-60 cells. The ability of p21 to be induced in a p53-independent manner during induction of differentiation in other systems and during development is now well established (BRUGAROLAS et al. 1995; DENG et al. 1995; HALEVY et al. 1995; MACLEOD et al. 1995; C. MISSERO et al. 1995; PARKER et al. 1995; ZHANG et al. 1995; LIU et al. 1996).

Induction of p21 is also apparent after induction of differentiation in additional p53-negative hematopoietic cell lines (STEINMAN et al. 1994; MACLEOD et al.

1995; ZHANG et al. 1995; LIU et al. 1996). These include an elevation in the levels of p21 after 24-h treatment of K562 cells with TPA resulting in megakaryocytic differentiation, of U937 cells with butyrate resulting in monocyte differentiation, and of M1 murine myeloid leukemia cells with interleukin (IL)-6 resulting in macrophage differentiation (STEINMAN et al. 1994). A potential role for p21 in normal bone marrow is also suggested by the induction of p21 following commitment to differentiation by treatment with granulocyte colony-stimulating factor (G-CSF) (STEINMAN et al. 1994). In contrast, treatment of HL-60 cells with G-CSF, which does not induce differentiation of HL-60 cells, does not induce p21 (STEINMAN et al. 1994).

An important question is whether induction of p21 is a consequence of growth suppression, or whether this gene expression change is a primary mediator of differentiation in leukemia cells. In mutant p53 murine erythroleukemia (MEL) cells, induction of p21 appears to correlate with induction of growth suppression that precedes terminal differentiation rather than maintenance of the growth-arrested phenotype (MACLEOD et al. 1995). p21 mRNA and protein levels increase 2 h after treatment with heamethylene bisacetamide (HMBA), whereas no significant change in p21 mRNA is apparent at later times following HMBA treatment when cells are terminally differentiated. A comparison of p21 mRNA (enhanced approximately two- to threefold) and protein levels (enhanced approximately tenfold) in HMBA-treated MEL cells suggests that p21 may also be regulated post-transcriptionally during HMBA-induced differentiation of MEL cells. Enhanced mRNA stability may also contribute to p21 upregulation during differentiation in human leukemia cells (Schwaller et al. 1995).

Variants of HL-60 cells displaying resistance to TPA-induced growth suppression and differentiation provide valuable experimental models to evaluate the potential involvement of p21 in growth arrest and terminal differentiation (MURAO et al. 1983; P.B. FISHER et al. 1984; JIANG et al. 1994b). Using HL-60 cells resistant to TPA-induced growth arrest and differentiation (HL-525 cells), a direct correlation is observed between the early induction of p21 and TPA- and RA-induced growth arrest and differentiation (JIANG et al. 1994b). In contrast, the effects of Vit D_3 on growth suppression and induction of differentiation are not directly correlated processes in HL-525 cells, whereas growth suppression and induction of differentiation are related changes in Vit D_3-treated HL-60 parental cells (JIANG et al. 1994b). In the HL-525 variant, growth is unaffected by treatment with 400 nM Vit D_3, whereas differentiation as monitored by the percentage of OKM1-positive cells is increased to a similar degree as in Vit D_3-treated, growth-arrested HL-60 cells. These results suggest that, in specific programs of differentiation in HL-525 cells, i.e., macrophage/monocyte differentiation induced by Vit D_3, p21 expression correlates directly with induction of differentiation in the absence of growth suppression. However, in most instances both differentiation and growth suppression occur simultaneously during differentiation in HL-60 and HL-525 variant cells.

To identify genes upregulated by Vit D_3, a cDNA library prepared from U937 cells was probed with RNA from either control or Vit D_3-treated cells (LIU et al.

1996). This differential screening strategy resulted in the identification and cloning of p21 (LIU et al. 1996). When U937 cells are treated with Vit D_3 for 4 h, expression of p21 and of additional INK4 CDI, including p15, p16, and p18, is upregulated. Evidence is presented that p21 is transcriptionally induced in U937 cells by Vit D_3 in a Vit D_3 receptor-dependent, but not p53-dependent manner. When transiently overexpressed in U937 cells, both p21 and p27 induce morphologic and antigenic markers of monocyte/macrophage differentiation in the absence of Vit D_3. In contrast, antigenic changes are not observed when amino-terminal deletions of p21 and p27, which remove the CDK-inhibitory activity of both proteins, are transiently expressed in U937 cells. These results suggest a potential causative role for inhibition of CDI, such as p21 and p27, in controlling terminal differentiation programs in U937 cells.

The studies briefly discussed above provide additional support for the wild-type p53-independent induction of p21 during the processes of growth arrest and differentiation. Furthermore, in specific cellular contexts, such as U937 cells, transiently expressing elevated levels of CDI, such as p21 and p27 can directly induce antigenic markers of monocyte/macrophage maturation and terminal differentiation in the absence of the cognate inducer Vit D_3. In contrast, in MEL cells elevated p21 expression does not correlate with the terminally differentiated state, and in HL-525 cells p21 expression correlates with Vit D_3-induced differentiation even in the absence of growth suppression. These experimental results indicate that p21 may play distinct and different roles in specific programs of hematopoietic growth and differentiation that are dependent upon undefined factors regulated in a cell-type dependent manner.

To directly define a possible role of p21 in the induction of differentiation and drug-mediated apoptosis in HL-60 cells, FREEMERMAN et al. (1996) used an antisense approach. HL-60 cells were produced that stably express p21 antisense, and these cells were compared with parental HL-60 cells with respect to their responses to differentiating and cytotoxic agents. Antisense-expressing HL-60 cells treated for 24 h with TPA displayed an attenuated induction of p21 compared with vector-transfected parental HL-60 cells. The modified response of the antisense p21-expressing HL-60 cells included a reduction in the percentage of cells undergoing G_1-phase arrest and expressing the monocytic maturation marker CD11b and an attenuation of the inhibition of cdc2 activity following TPA treatment. In contrast, no differences were observed in the phosphorylation status of Rb, in E2F complex formation, or in p27 induction following TPA treatment. However, TPA did reduce the cloning efficiency of vector control cells to a greater extent than it did that of antisense expressing-HL-60 cells. In contrast, treatment of the two cell types with the antimetabolite 1-β-D-arabinofuranosylcytosine (ara-C) resulted in an equal susceptibility to G_1-phase arrest, apoptosis, and inhibition of clonogenicity, processes that are not associated with p21 and p27 induction or Rb phsophorylation in HL-60 cells. These findings demonstrate that dysregulation of p21 in p53-null HL-60 cells interferes with TPA-related G_1-phase arrest, CDK-2 inhibition, differentiation, and loss of clonogenic survival in the absence of obvious changes in Rb phsophorylation status or E2F complex formation. This data also documents that p21 induction does not appear to be necessary for ara-C-induced apoptosis, G_1-phase arrest, or the resulting reduction in the self-renewal capacity of HL-60 cells.

2.2.3 Muscle Cell Differentiation

Cell culture model systems are providing important insights into the biochemical and molecular determinants of terminal skeletal muscle differentiation (WEINTRAUB 1993; LASSAR et al. 1994; OLSON and KLEIN 1994; WALSH et al. 1996). When grown in medium lacking serum, cultured myoblasts fuse to form multinucleated myotubes, activate specific programs of gene expression, and exit irreversibly from the cell cycle (WEINTRAUB 1993; LASSAR et al. 1994; OLSON and KLEIN 1994; WALSH et al. 1996). Induction of differentiation results from the expression of the myogenic basic helix–loop–helix (bHLH) transcription factors, including the myogenic determination proteins MyoD and Myf-5 (WEINTRAUB 1993; LASSAR et al. 1994; OLSON and KLEIN 1994; WALSH et al. 1996). Although myoblasts are predetermined to the muscle lineage, they are prevented from completing differentiation by the presence of Id, a dominant-negative HLH factor that inhibits the activity of the bHLH factors (BENEZRA et al. 1990). Id can form complexes with MyoD and Myf-5, nullifying their ability to bind to and activate downstream genes involved in mediating differentiation (BENEZRA et al. 1990). Growing myoblasts in medium devoid of mitogens decreases the levels of Id, thereby allowing the interaction of myogenic bHLH factors with the abundant E12 and E47 proteins. Formation of these complexes results in the formation of myotubes that cannot undergo mitosis.

An important regulator of myoblast differentiation is the Rb tumor suppressor gene product (COPPOLA et al. 1990; Y. GU et al. 1993; CARUSO et al. 1993). In differentiated skeletal muscle, Rb is present predominantly in the hypophosphorylated (active) form, and loss of Rb resulting from gene inactivation blocks the normal G_0-phase arrest found in differentiated skeletal muscle (SCHNEIDER et al. 1995). In the hypophosphorylated state, Rb suppresses the transcriptional activity of E2F that is needed for the expression of S-phase genes involved in cell cycle progression (NEVINS 1992; SHERR 1993; HARTWELL and KASTAN 1994; HUNTER and PINES 1994; CHELLAPPAN 1994; HOROWITZ and UDVADIA 1995; WEINBERG 1995).

A number of recent studies demonstrate a relationship between p21 expression and muscle terminal differentiation (GUO et al. 1995; HALEVY et al. 1995; C. MISSERO et al. 1995; PARKER et al. 1995; SKAPEK et al. 1995). During skeletal muscle differentiation, p21 expression is markedly induced, resulting in increased p21 mRNA and protein (GUO et al. 1995; HALEVY et al. 1995; PARKER et al. 1995; SKAPEK et al. 1995; WANG and WALSH 1996). p21 is found in immunoprecipitable complexes with the CDK in myotubes, and this CDK-inhibitory activity persists when myotubes are recultured in mitogen-rich medium (GUO et al. 1995; ANDRES and WALSH 1996). In this context, the high levels of p21 expression in myotubes may directly contribute to the irreversibility of the differentiated state in fused myoblasts. Overexpression of cyclin D_1, but not cyclins A, B, D_2, D_3, or E, can inhibit myogenesis as monitored by the ability of MyoD to transcriptionally activate the muscle-specific creatine kinase (MCK) promoter and muscle-specific genes

(RAO et al. 1994; SKAPEK et al. 1995). Transfection of myoblasts with p21 and p16 reverses cyclin D_1 inhibition of MyoD and augments muscle-specific gene expression in cells maintained in high concentrations of serum (SKAPEK et al. 1995). These results suggest that active cyclin–CDK complexes suppress MyoD function in proliferating cells. When MyoD is transfected into either murine myocytes or nonmyogenic cells, including mouse 10T1/2, monkey CV1, and human osteosarcoma U20S, p21 is induced (HALEVY et al. 1995; PARKER et al. 1995). In addition, MyoD induces p21 in p53-deficient mouse embryo fibroblasts and myogenic cells lacking MyoD and myogenin, suggesting that these factors are not required for p21 induction (HALEVY et al. 1995; PARKER et al. 1995). Moreover, the expression of mouse p21 correlates with terminal differentiation of multiple cell lineages, including skeletal muscle, cartilage, skin, and nasal epithelium in a p53-independent manner. Taken together, these results suggest that p21 may function as an inducible growth inhibitor during development and that this protein may be part of a positive-regulatory loop that functions to promote the differentiated state in skeletal muscle (PARKER et al. 1995; WALSH et al. 1996).

Recent studies suggest that the CDI p21 and p16 may provide a protective role in muscle differentiation by preventing apoptosis (WANG and WALSH 1996). When proliferating murine C2C12 myoblasts undergo mitogen deprivation, they can either terminally differentiate or undergo programmed cell death (apoptosis). Differentiated myotubes are resistant and myoblasts are sensitive to mitogen withdrawal-induced apoptosis. Ectopic expression of p21 or p16 blocks apoptosis during myocyte differentiation. These results suggest that the induction of specific CDI may function to prevent differentiating myocytes from undergoing programmed cell death and may contribute to establishment of the postmitotic state (WANG and WALSH 1996). Further studies are necessary to determine whether this phenomenon is restricted to muscle differentiation or whether it represents a general mechanism by which CDI function in terminal differentiation in other cell lineages.

2.2.4 Keratinocyte Differentiation

Keratinocyte differentiation represents a well-characterized model system for analyzing the process of differentiation and cellular maturation as cells migrate toward the outermost layer of the epidermis and ultimately exfoliate (HENNINGS et al. 1980; FILVAROFF et al. 1990; DOVER and WRIGHT 1991; CALAUTTI et al. 1995; C. MISSERO et al. 1995; MISSERO and DOTTO 1996). The addition of calcium to mouse primary keratinocytes induces growth arrest and a terminal differentiation program that reflects the process normally observed in the external layers of the skin (HENNINGS et al. 1980). In contrast, growth of mouse primary keratinocytes in transforming growth factor (TGF)-β inhibits growth without inducing markers of differentiation (C. MISSERO et al. 1995). In these experimental situations, p21 is induced by 4 h after removal of calcium and progressively increases up to 18–24 h, whereas no consistent induction of p21 is apparent during the same period when primary mouse keratinocytes are treated with TGF-β.

As observed in human melanoma cells induced to terminally differentiate by treatment with IFN-β + MEZ (JIANG et al. 1995b, 1996a), p53 protein levels decrease in primary keratinocytes induced to differentiate with calcium (C. MISSERO et al. 1995; MISSERO and DATTO 1996). Using primary mouse keratinocytes derived from p53-null mice, induction of differentiation by the addition of calcium increases p21 mRNA levels (C. MISSERO et al. 1995). Moreover, transfection of normal and p53-null mouse keratinocytes with a 2.4-kb DNA fragment of the p21 promoter linked to a luciferase gene, with or without the p53 recognition motif at its 5'-end, results in enhanced expression upon induction of differentiation (C. MISSERO et al. 1995). These findings demonstrate that induction of keratinocyte differentiation by calcium correlates with an increase in p21 promoter activity that is independent of p53 expression.

A potential relationship between expression of the adenovirus E1A-associated p300 protein and a differentiation signal linked to growth arrest has been provided using the mouse keratinocyte model (C. MISSERO et al. 1995; MISSERO and DOTTO 1996). In this system, the E1A protein inhibits the activity of the p21 promoter following calcium addition. However, when a wild-type p300 or a mutated form of p300 lacking the E1A-binding region is added in combination with E1A, suppression of the calcium responsiveness of the p21 promoter does not occur (C. MISSERO et al. 1995). Further support for an association between p300 expression and keratinocyte differentiation is provided using the involucrin promoter, which is necessary for keratinocyte differentiation and requires p300 for induction. Moreover, the human p300 has been found to bind both in vivo and in vitro to MyoD and to potentiate MyoD-dependent transactivation (YUAN et al. 1996). Additional studies are necessary to determine whether p300 also contributes to other programs of differentiation.

2.2.5 Development

Important questions relate to the role of p21 in normal development and cancer. To address this issue, DENG et al. (1995) produced p21-null ($p21^{-/-}$) mice by homologous recombination, and BRUGAROLAS et al. (1995) developed chimeric mice composed partly of $p21^{-/-}$ and partly of $p21^{+/+}$ cells. On the basis of p21 induction by wild-type p53 (EL-DEIRY et al. 1993), it was assumed that altering p21 levels would directly impact on specific phenotypes normally regulated by wild-type p53, such as cell cycle progression, differentiation, development, growth properties, tumorigenesis, apoptosis, and mitotic spindle checkpoint. An important function of wild-type p53 is maintenance of cell cycle progression by the control of specific checkpoints (HARTWELL and KASTAN 1994). Checkpoints are also important in maintaining cellular DNA integrity and regulating normal cell cycle transitions. When the DNA of cells containing wild-type p53 are damaged, a coordinate arrest of cell cycle occurs and genes assisting in and mediating DNA repair are induced. In this manner, cells are arrested in G_1 phase, permitting repair of the DNA damage and preventing replication and propagation of defective chromosomes. In mice embryo fibroblasts deficient in p53, no G_1-phase, checkpoint is evident, whereas this checkpoint is apparent in mice embryo fibroblasts containing wild-type p53. Analysis of this phenomenon in p21$^{-/-}$ mouse embryo fibroblasts indicates a significant impairment in the G$_1$-phase response to DNA damage and nucleotide pool alterations (DENG et al. 1995; Brugarolas et al. 1995). These findings directly establish a role for p21 in the G$_1$-phase checkpoint.

In contrast to the predictable effect of the absence of p21 on cell cycle progression following DNA damage, many of the anticipated aberrations in p21$^{-/-}$ mice are not apparent. For example, in contrast to p53$^{-/-}$ mice that acquire spontaneous malignancies by 6 months of age (DONEHOWER et al. 1992; HARVEY et al. 1994; JACKS et al. 1994; WILLIAMS et al. 1994b), p21$^{-/-}$ mice develop normally and do not show signs of early tumor development (DENG et al. 1995). Although p21 appears to directly contribute to differentiation in several cell culture and in vivo model systems, tissue development and differentiation are normal in p21$^{-/-}$ mice and the migration-associated differentiation of the four epithelial adult lineages in adult small intestine are unaltered in p21$^{-/-}$ and p21$^{+/+}$ chimeric mice (BRUGAROLAS et al. 1995). Moreover, p21 is not required for the p53-dependent mitotic spindle checkpoint or for p53-dependent apoptosis (DENG et al. 1995; BRUGAROLAS et al. 1995). These findings suggest that p21 may be redundant for normal development and terminal differentiation. Moreover, wild-type p53 may induce additional downstream genes that are necessary to mediate specific programs of p53-dependent cellular responses.

An alternate methodology to evaluate the functional role of p21 in vivo is to use tissue-specific expression in transgenic animals (WU et al. 1996). Targeted expression of p21 in hepatocytes in transgenic mice results in an inhibition in hepatocyte proliferation. A consequence of this disturbance is a decrease in the total number of mature hepatocytes, resulting in abnormal tissue organization, runted liver, suppressed body growth, and elevated mortality. The transgenic p21 is found predominantly associated with cyclin D$_1$–CDK4 in liver, emphasizing the potential importance of this interaction in normal liver development. Evidence is provided that p21 may also cause growth arrest in G$_2$ phase of the cell cycle in specific hepatocytes. In contrast to normal animals, partial hepatectomy fails to induce hepatocyte growth in p21 transgenic animals. These studies indicate that alterations in p21 in specific target organs can affect normal development. Further studies are necessary to determine whether elevated p21 targeted to other tissues in transgenic animals will affect their developmental program.

2.2.6 Mode of Action

Most of the CDK in normal fibroblasts are organized into quaternary complexes consisting of a CDK, a cyclin, proliferating cell nuclear antigen (PCNA), and p21 (XIONG et al. 1992; ZHANG et al. 1993). In nontransformed cells, p21 binds directly to several CDK, including cdc2 (cdk1), cdk2, and cdk4 (HARPER et al. 1993; XIONG et al. 1993a, b; ZHANG et al. 1993; DULIC et al. 1994). In these cells, the CDK coprecipitate forming enzymatically active complexes with several cyclins, including cyclins A, B, D, and E (XIONG et al. 1993b; ZHANG et al. 1993, 1994). A number of studies document that p21 is a universal inhibitor of CDK enzyme activity with the ability to induce cell cycle arrest in G_1 phase of the cell cycle (EL-DEIRY et al. 1993; Y. GU et al. 1993; HARPER et al. 1993; XIONG et al. 1993a; NODA et al. 1994).

The presence of p21 in enzymatically active quaternary complexes in normal proliferating fibroblast cells represents a paradox. Similarly, the induction of p21 in quiescent fibroblasts and T lymphocytes induced to proliferate by mitogenic factors (FIRPO et al. 1994; Y. LI et al. 1994a; NODA et al. 1994; SHEIKH et al. 1994) is at odds with the suggestion that p21 is a universal inhibitor of CDK. This conundrum is resolved by the demonstration that cyclin kinases containing p21 can exist in both enzymatically active and inactive states depending on the concentration of p21 (ZHANG et al. 1994). Evidence is presented that enzymes containing p21 can change between active and inactive states, probably through changes in the stoichiometry of the p21 subunits (ZHANG et al. 1994). When the complex contains a single p21 molecule, it is enzymatically active, whereas inactive complexes contain multiple p21 subunits.

The presence of PCNA in the quaternary complex with p21 suggests a potential relationship between p21 expression and the processes of DNA replication and/or DNA damage repair. PCNA is essential for DNA replication and nucleotide-excision repair. p21 can directly bind to PCNA and inhibit the in vitro replication of simian virus 40 (SV40) DNA, whereas it does not inhibit PCNA-dependent nucleotide-excision repair (WAGA et al. 1994; R. LI et al. 1994; SHIVJI et al. 1994). The capacity of p21 to directly inhibit PCNA-dependent DNA replication, but not alter DNA repair, is compatible with the hypothesis that genetic damage can lead to inactivation of chromosomal replication while allowing damage-responsive repair (R. LI et al. 1994).

The p21 protein can directly inhibit the cyclin-CDK family of kinases and cell growth and can bind to PCNA, resulting in an inhibition of in vitro PCNA-dependent DNA replication (EL-DEIRY et al. 1993; WAGA et al. 1994; FLORES-ROZAS et al. 1994). These results suggest that the p21 protein may display multiple properties that could occur as a result of binding of these different molecules to the same region of this protein or as a consequence of differential binding to distinct domains in this protein. This question has been addressed using biochemical and molecular approaches indicating that the growth- and CDI-inhibitory domains of p21 reside in the amino-terminal domain of this protein, whereas the PCNA-binding and -inhibitory activities reside in the carboxy-terminal domain (LUO et al. 1995; ZAKUT and GIVOL, 1995; J. CHEN et al. 1995b). These experiments indicate that the domains of the p21 protein that mediate cyclin-CDK and PCNA interactions are distinct, supporting the idea that p21 has dual target molecules and multiple functions in cells.

Experiments have also been performed to define the regions of p21 responsible for inducing growth arrest and for inhibiting DNA replication in vivo. In studies employing *Xenopus* extracts, evidence is presented that p21 inhibits replication of double-stranded DNA through its CDK-inhibitory domain by selectively targeting cyclin E-cdk2 complexes as opposed to PCNA (X. CHEN et al. 1995a). When transfected into p53-negative Saos-2 cells, only the C-terminal domain of p21 inhibited colony formation (X. CHEN et al. 1995a). In contrast, using transient transfection assays with constructs permitting nuclear localization (SV40 nuclear localization signal), both N- and C-terminal domains of p21 suppressed cellular DNA synthesis in mink lung epithelial cells (R-1B/L17), with the N-terminal domain being somewhat more efficient than the C-terminal domain (LUO et al. 1995). In p53-negative human fibroblasts, deletion of amino acids 66–71 from the N-terminal region of p21 eliminates cdk2-inhibitory activity in vitro with a retention of DNA synthesis-inhibitory activity (NAKANISHI et al. 1995). These studies indicate that, although it is less effective, the PCNA-binding domain of p21 is growth inhibitory and the ability of p21 (N- or C-terminal regions) to block cell cycle progression is concentration dependent.

A potential role for cyclin D_1 in p21-mediated growth suppression is also indicated (X. CHEN et al. 1995a; DEL SAL et al. 1996). Increases in wild-type p53 protein levels result in increased p21 and cyclin D_1 expression. In addition, induction of cyclin D_1 can also be mediated by p21 (X. CHEN et al. 1995a). The elevated levels of cyclin D_1 in murine cells containing a temperature-sensitive wild-type p53 gene are a consequence of both post-transcriptional and post-translational regulatory mechanisms (DEL SAL et al. 1996). Using an antisense cyclin D_1 cDNA strategy, evidence is presented that cyclin D_1 levels contribute to wild-type p53 G_1-phase arrest in murine cells (DEL SAL et al. 1996). Although antisense suppression of either cyclin D_1 or p21 can retard the onset of p53-induced growth arrest, elimination of either of these gene products does not override a preexisting p53-induced G_1-phase block in murine cells. These studies implicate both cyclin D_1 and p21 in p53-induced growth arrest in specific target cells. A potential involvement of cyclin D_1 in myogenesis has been demonstrated (RAO et al. 1994; SKAPEK et al. 1995). It remains to be determined whether cyclin D_1 is also a major contributor to additional programs of differentiation.

Recent studies by HERMEKING et al. (1995) present evidence for an interesting and novel mechanism of cell cycle regulation involving a putative inhibitor of CDI. c-myc can inhibit p53-induced G_1-phase arrest in mouse fibroblast cells containing a temperature-sensitive p53 gene (p53^{val135}) without increasing the expression of CDK or cyclins normally involved in G_1/S-phase transition. This blocking of G_1-phase arrest by c-myc occurs because this proto-oncogene inhibits the action of p21 on the CDK–cyclin complexes by inducing a heat-labile inhibitor of p21. Further studies are required to determine whether the same or a similar inhibitor contributes to cellular differentiation.

3 Cyclin-Dependent Kinase Inhibitor p27^{KIP1}

p27 was first identified as an activity that was capable of mediating TGF-β, and contact inhibition induced growth arrest in mink epithelial cells (POLYAK et al. 1994a). Current evidence suggests that TGF-β arrests cell proliferation by preventing progression at the late G_1 phase of the cell cycle (LAIHO et al. 1990). Following release of Mv 1 Lu mink lung epithelial cells from serum starvation, cultures progress through the cell cycle normally, and phosphorylated forms of Rb protein appear after 12 h. When TGF-β is added to the growth medium 6 h after serum stimulation, the cells are arrested in late G_1-phase and the Rb protein remains in an unphosphorylated state (KOFF et al. 1993). This lead to the hypothesis that TGF-β blocks a G_1-phase CDK activity by an unknown mechanism (Fig. 1). Further studies revealed that, although TGF-β-treated cells have comparable amounts of cyclin E and cdk2 as untreated cells, there is no kinase activity associated with these molecules. This raised the possibility that TGF-β either prevents formation of cyclin E–cdk2 complexes or inhibits the kinase activity of the associated complexes. Extracts prepared from cells treated with TGF-β prevent the activation of endogenous CDK by exogenously added cyclins, suggesting that these extracts may contain an inhibitor preventing activation (POLYAK et al. 1994a).

The notion of an inhibitor of cyclin activity in TGF-β-treated cells was confirmed by the observation that addition of cyclin E to an extract from proliferating cells produces a linear increase in cyclin E-associated activity, whereas addition of as much as three times the physiological concentration of cyclin E to TGF-β-treated cells results in no increase in immunoprecipitable cyclin E-associated histone H_1 kinase activity (KOFF et al. 1993). Addition of a large excess of cyclin E results in detectable amounts of cyclin E-associated kinase activity. This finding, along with the fact that the levels of cyclin E and CDK2 are similar in the TGF-ß-treated and untreated extracts, suggests that there is indeed a titratable inhibitor of cyclin E, associated kinase activity in these extracts (KOFF et al. 1993). The inhibitor of the cyclin E associated kinase activity associates with cyclin E–CDK2 complexes and cannot be detected with either protein alone. The inhibitor is able to bind to a mutant cdk2 lacking kinase activity, and reversal of the inhibition by the addition of cyclin E–cdk2 does not involve the phosphorylation of the inhibitor, but rather a direct titration of this factor. The inhibitory activity can also be removed from the extracts following incubation with cyclin D2–CDK4, suggesting that this inhibitor is capable of targeting different G_1-phase cyclins and associated kinases (POLYAK et al. 1994a). Attempts were made to purify the inhibitory activity using cyclin E–CDK2 affinity columns, resulting in the identification of a 27-ka protein present in extracts from TGF-β-treated cells, but absent in cells in late G_1 phase. This activity is surprisingly heat stable, and brief heat treatment at 100°C does not abolish the ability of p27 to bind to cyclin E–cdk2 or alter its inhibitory activity (POLYAK et al. 1994a).

p27 was cloned by two groups using different approaches (POLYAK et al. 1994b; TOYOSHIMA and HUNTER 1994). In one approach, p27 was affinity purified on a cyclin E–cdk2 column and microsequenced, and the partial sequence was used to screen a mink Mv 1 Lu library to obtain the full-length clone (POLYAK et al. 1994b). p27 was also identified using a yeast two-hybrid system, with cyclin D–cdk4 as the bait (TOYOSHIMA and HUNTER 1994). The genes obtained by the different methods were identical, and the protein was found to be similar to the p21 protein, indicating the existence of a family of CDI proteins. The predicted human, mouse, and mink p27 proteins are highly related, showing approximately 90% identity. There is 44% identity to the p21 protein, but it is mainly confined to a 60-amino acid region in the amino-terminal portion of the protein. The p27 protein contains a bipartite nuclear localization signal and a consensus cdk2 phosphorylation site at the C-terminal end (POLYAK et al. 1994b). Chromosome mapping studies indicate that p27 is located on chromosome 12q13. Chronic lymphoid leukemia is frequently associated with chromosomal abnormalities at chromosome 12. Although no genomic alterations of p27 have been identified to date, the potential role of this CDI gene in leukemogenesis is under active investigation (BULLRICH et al. 1995; PIETENPOL et al. 1995; PONCE-CASTANEDA et al. 1995).

The region of p27 involved in inhibiting kinase activity resides at the amino terminus between residues 28 and 79, a region with extensive similarity to the p21 protein (POLYAK et al. 1994b). A recombinant peptide corresponding to this region was able to inhibit phosphorylation of Rb by cyclin A–cdk2, cyclin E–cdk2, and cyclin D2–cdk4, although the latter two were less susceptible to inhibition by the peptide than the full-length protein. Deletion of three amino acids at the amino terminus of the peptide reduces its effectiveness, and deletion of seven amino acids completely abolishes its inhibitory activity (POLYAK et al. 1994b).

The ability of p27 to bind to different cyclin–CDK complexes has been investigated. In one study, p27 inhibited cyclin E–cdk2 with maximum efficiency; an eightfold higher concentration was required to inhibit cyclin A–cdk2 to a similar extent, and this concentration was not sufficient to block cyclin B–cdc2 (POLYAK et al. 1994b). In a related set of experiments, the binding of p27 to various cyclins and CDK was assessed by glutathione S-transferase (GST)-affinity chromatography (TOYOSHIMA and HUNTER 1994). In these studies, it was found that the murine p27 bound with maximum affinity to cyclins D_1, D_2, and D_3. In contrast, cyclin E bound only weakly, and cyclin A did not bind at all. A co-immunoprecipitation assay on mouse 3T3 extracts indicates that cyclin D_1 is detectable in p27-immune precipitates; approximately 50% of the cellular cyclin D_1 is estimated to be associated with p27. Surprisingly, in these extracts, other cyclins are not detected in p27-immune precipitates (TOYOSHIMA and HUNTER 1994).

The efficacy of p27 in arresting cell proliferation was assessed by transfection assays using two cell lines, Mv 1 Lu and Saos-2. In Mv 1 Lu cells, transfection of p27 results in a decrease in the percentage of cells in S phase from 35% to 7% (POLYAK et al. 1994b). This correlates with a substantial reduction in ^{125}I-labeled deoxyuridine incorporation in the transfected cells. Similar results are obtained with Saos-2 cells; the percentage of cells in the S phase drops from 32.7% to 11.5% (TOYOSHIMA and HUNTER 1994). There is a corresponding increase in the percentage of cells in G_1 phase, from 46.2% to 76%. Both groups have convincingly demonstrated that overexpression of p27 can lead to a block in the cell cycle, preventing progression from G_1 to S phase.

The expression pattern of p27 in human tissues reveals a 2.5-kb message that is present in comparable amounts in most tissues (POLYAK et al. 1994b; TOYOSHIMA and HUNTER 1994). Surprisingly, there are no major changes in the levels of p27 message in growth-arrested, dividing, or TGF-β-treated mink epithelial cells.

Similarly, the levels of p27 protein remain constant in 3T3 cells after serum starvation and stimulation or release from a nocodazole block (TOYOSHIMA and HUNTER 1994). These results suggest that p27 is functionally regulated at a post-translational level by mechanisms that are not yet clear.

3.1 Regulation of p27 Levels in Cells

Recent studies suggest that the accumulation of p27 during the cell cycle is regulated, at least in many instances, at the level of translation. Arrest of HeLa cells by ellovastatin leads to a modest increase in the levels of p27 protein, whereas the levels of the message remain constant (HENGST and REED 1996). More profound changes are observed when human fibroblasts are starved of serum and then released from starvation. When HL-60 cells are treated with Vit D_3 for 72–120 h, there is a considerable accumulation of p27 protein, but the mRNA levels remain unchanged (HENGST and REED 1996). Pulse-chase experiments indicate that the half-life of the protein can also vary, and the protein is more stable in growth-arrested cells. It is possible that a rapid increase in the translation of p27 may contribute to the negative regulation of G_1 phase, and altered protein half-life may be important for maintenance of the growth-arrested state.

Post-translational control of p27 is also suggested by the observation that the level of this CDI is regulated by proteolysis through the ubiquitin proteasome pathway (PAGANO et al. 1995). Addition of inhibitors of the chymotryptic site of the proteasome causes p27 to be ubiquitinated. p27 is also ubiquitinated in vitro, suggesting that this may be an important level of p27 regulation. The human ubiquitin-conjugating enzymes Ubc2 and Ubc3 are specifically involved in the ubiquitination of p27 (PAGANO et al. 1995). Interestingly, quiescent cells have lower amounts of ubiquitinating activity compared to proliferating cells, suggesting that this could contribute, at least in part, to the longer half-life of p27 in quiescent cells. Therefore, it appears that the relative abundance of p27 is modulated by a combination of translational and post-translational mechanisms. It is likely that a combination of these two modes of regulation leads to the maintenance of a precise balance between the relative amounts of p27 and cyclin–CDK; this could be altered in favor of growth arrest or proliferation upon receiving the appropriate mitogenic/inhibitory signals (PAGANO et al. 1995).

3.2 Role in Proliferation and Differentiation

As a CDI inhibitor, it was anticipated that p27 would be a key component in cell cycle control, a prediction that has been validated by recent studies (Fig. 1). Control of the G_1/S-phase restriction point requires p27 (COATS et al. 1996). The appearance of p27 message following mitogen stimulation coincides with the inactivation of cyclin–CDK complexes, and a downregulation of p27 is necessary for mitogens to effect cell cycle progression. Depletion of p27 by antisense oligonucleotides abrogates the ability of cells to undergo growth arrest following mitogen depletion (COATS et al. 1996).

The p27 protein is implicated in growth arrest mediated by TGF-β, cyclic adenosine monophosphate (cAMP), and rapamycin. As described above, studies on growth arrest induced by TGF-β resulted in the characterization and cloning of p27. A similar relationship between p27 levels and cAMP-mediated growth arrest of macrophages has been demonstrated (KATO et al. 1994a). Macrophages growing in complete growth medium distribute through the different phases of the cell cycle in a similar manner as cell lines maintained in culture. Treatment with CSF-1 in combination with analogues or inducers of cAMP arrest macrophages predominantly in G_1 phase and the cells remain arrested in G_2 phase. Macrophages deprived of CSF-1 are arrested in early G_1 phase, and addition of CSF-1 allows them to reenter the cell cycle (MATSUSHIME et al. 1991). Addition of rapamycin or 8Br-cAMP within 5 h of CSF-1 addition prevents cells from entering S phase, suggesting that the cells are arrested in mid-G_1 phase, prior to the late G_1-phase restriction point. This time point corresponds to the first appearance of cyclin D-associated kinase activity (MATSUSHIME et al. 1991). Both 8Br-cAMP and rapamycin repress cyclin D-mediated phosphorylation of the Rb protein, preventing the cells from progressing to S-phase. The levels of cdk4 and cyclin D remain unchanged in cells treated with CSF-1 plus 8Br-cAMP, Bt2cAMP, or rapamycin, suggesting that each compound inhibits the phosphorylation of Rb by blocking the kinase activity of the assembled holoenzyme.

Experiments to evaluate whether any of the three known inhibitors of cdk4 activity, i.e., p21, p16, or p27, are involved in the cAMP-mediated repression reveal that p27 is most probably responsible for the inhibition (KATO et al 1994a). This conclusion is based on the fact that the p16 gene is deleted in the BAC1.2F5A cell line used for the study, and three- to fourfold more p27 is detected in cyclin D-immune precipitates obtained from 8Br-cAMP-treated cells than from those treated with CSF-1 alone. No increase is detected in cells treated with rapamycin. p27 inhibits cyclin D–cdk4 activity by binding to the holoenzyme and preventing the activating phosphorylation of threonine-172 by the CAK (GU et al. 1992). Immunodepletion of p27 from cAMP-treated cell extracts can reverse the inhibition of cyclin D–cdk4 activity, accompanied by an increase in CAK-mediated phosphorylation of cdk4 (KATO et al. 1994a).

Since p27 acts stoichiometrically, the two possible ways that p27 can inhibit cyclin D–cdk4 activity include a reduction in the levels of cyclin D–cdk4 protein or an increase in the amount of p27. The latter mechanism appears correct, since there is a two- to threefold increase in the levels of p27 in quiescent cells treated with CSF-1 plus 8Br-cAMP compared to cells treated with CSF-1 alone (KATO et al. 1994a). This suggests that cAMP addition raises the p27 threshold, resulting in increased binding to cyclin D–cdk4, inhibition of CAK-mediated activation, and concomitant inhibition of cyclin D–cdk4 kinase activity.

A similar role for p27 is suggested in T cell mitogenesis. During T cell mitogenesis, antigen receptor signaling promotes synthesis of cyclin E and cdk2, and IL-2 promotes cyclin E–cdk2-associated kinase activity (FIRPO et al. 1994). IL-2 promotes the activation of cyclin E–cdk2 by causing the elimination of p27 protein, and this inhibition is reversed by rapamycin, a strong immunosuppressant. This conclusion is supported by the observation that cyclin E–cdk2 complexes in cells treated with IL-2 have lower amounts of p27 associated with them than in cells treated with IL-2 plus rapamycin (NOURSE et al. 1994). Antibodies specific for p27 eliminate the inhibitory activity in IL-2-starved and rapamycin-treated cells, suggesting that elevated levels of p27 account for all of the inhibitory activity detected in growth-arrested cells. Immunoblotting of whole-cell extracts demonstrates that IL-2 induction gradually reduces the amount of p27, and rapamycin prevents the removal of p27 (NOURSE et al. 1994). A low level of p27 is detectable in cells treated with IL-2, and they progress through S phase. This finding suggests that p27 levels function at a threshold and that this threshold level can be modulated by IL-2 or rapamycin to bring about the observed effects on proliferation. This study did not provide any evidence for p27 blocking the activation of cdk2 by CAK (NOURSE et al. 1994), as in the case of cdk4 (KATO et al. 1994a; APPRELIKOVA et al. 1995). This conclusion is based on the fact that the Thr-160-phosphorylated form of cdk2 is present in comparable amounts in rapamycin-blocked or proliferating cells. This observation suggests that p27 blocks the activity of cyclin E–cdk2 that has already been phosphorylated by the CAK enzyme. It is also clear that repression of p27 in T lymphocytes by extracellular mitogens is not restricted to this agent, since serum growth factors can elicit the same effect.

p27 redistributes itself between different cyclin–CDK complexes during the cell cycle and following growth arrest by lovastatin, UV irradiation, or growth factor deprivation (POON et al. 1995). In cells arrested by growth factor depletion, most of the p27 occurs in monomers; however, after entering G_1 phase, most of the p27 is associated with cyclin D–cdk4 complexes. As the cells progress to S phase, p27 is redistributed to cyclin A–cdk2 complexes. In most experiments, the interaction of p27 with cyclin E–cdk2 complexes has not been determined. It is hypothesized that cyclin D1–cdk4 acts as a reservoir for p27 and that p27 is redistributed to cyclin A–cdk2 complexes when the cells enter S phase, or even when cells are growth arrested by growth factor depletion, lovastatin treatment, UV irradiation, or by other means (POON et al. 1995). Of note is the observation that the adenovirus E1A protein, which can reverse TGF-β-mediated growth arrest, targets p27 (MAL et al. 1996). E1A can bind to p27 and inhibit its function, providing the first example of a viral oncoprotein targeting a CDI (MAL et al. 1996). This situation is similar to that observed with the adenovirus E1A and similar oncoproteins that inactivate the Rb family of tumor suppressors, which are also important negative regulators of the cell cycle.

The suggestion that there is a redistribution of p27 between different cyclin–CDK complexes has been confirmed by elegant experiments on TGF-β-mediated growth arrest of Mv 1 Lu cells (REYNISDOTTIR et al. 1995). Proliferating Mv 1 Lu cells and human keratinocytes contain high levels of p27, distributed in complexes of cyclins with cdk2, cdk4, and cdk6. Treatment of cells with TGF-β induces p15INK4B expression, which is specific for cdk4/cdk6, and this induction coincides with a release of p27 from cyclin D-cdk4/cdk6 complexes (REYNISDOTTIR et al.

1995). This leads to an increased association of p27 with complexes containing cdk2. Furthermore, recombinant p15 inhibits the binding of p27 to cyclin D–cdk4/cdk6 complexes, which supports the suggestion that elevated levels of p15 cause the release of p27 from cdk4/cdk6 complexes upon TGF-β treatment. Even though TGF-β also induces p21, it does not cause the release of p27 from cdk4/cdk6 complexes. These findings suggest that TGF-β causes a G_1-phase arrest by inducing the expression of p15 as well as p21, and that p15 induction causes an inhibition of cyclin D–cdk4/cdk6 activities (REYNISDOTTIR et al. 1995). Even the kinase activities of complexes containing cdk2 are inhibited due to the release of p27 from cdk4/cdk6 kinase as well as the induction of p21. It appears that TGF-β-mediated growth arrest is regulated by multiple CDI.

p27 can affect brain-specific cdk5 depending on the activator associated with cdk5 (M.H. LEE et al. 1996). The levels of p27 are quite high in post-mitotic neurons compared to their progenitor neuroblasts, and p27 inhibits cdk2 in post mitotic neurons. Surprisingly, cdk5, which predominantly associates with the non-cyclin activator p35 in neurons, is resistant to inhibition by p27. cdk5 can be inhibited by p27 when it is associated with cyclin D2. p27 can recognize, bind, and inhibit only the cyclin D2–cdk5 complex in vitro and in vivo, but not a p35–cdk5 complex (M.H. LEE et al. 1996). It is possible that cdk5 remains active in post-mitotic neurons in spite of high amounts of p27 by virtue of the ability of cdk5 to associate and function in conjunction with p35, thus precluding inhibition.

3.3 Role in Development

The CDK and molecules that affect their function play a major role in cell proliferation, differentiation, and development. It has previously been demonstrated that molecules involved in cell cycle regulation, such as the Rb, p107, and p130 proteins, play major roles in embryogenesis and development. Attempts have been made to unravel the role of these proteins during development by establishing whether any particular proteins are important for the successful completion of specific developmental pathways. A commonly used strategy to define these relationships is to construct mutant mice with a homozygous deletion of the gene of interest.

The role of p27 in murine development has been assessed by three groups by generating mice lacking both alleles of the p27 gene (NAKAYAMA et al. 1996; FERO et al. 1996; KIYOKAWA et al. 1996). Two groups engineered the deletion of the complete coding sequence of p27 from the mice (NAKAYAMA et al. 1996; FERO et al. 1996), and another group replaced the functional p27 gene in the mouse genome with a nonfunctional N-terminal deletion mutant (KIYOKAWA et al. 1996). The most notable phenotype is an increased overall size of the resulting mice, which indicates a potential role for p27 in restricting cell proliferation during the course of development. This finding is in agreement with the known function of p27 in cell cycle arrest. The major phenotypic characteristics of the p27 null mice are described below.

3.3.1 General Phenotype of p27-Null Mice

Mating of heterozygous mice lacking one allele of the wild-type p27 gene produced mice that were wild type, heterozygous, and nullizygous for the p27 gene in approximately mendelian proportions. This suggests that there is no embryonic lethality associated with p27 knockout, and early developmental pathways are not greatly impaired. At birth, the homozygotes are indistinguishable by size or other phenotypic features from their wild-type and heterozygote littermates (NAKAYAMA et al. 1996; KIYOKAWA et al. 1996; FERO et al. 1996).

The majority of mice lacking both copies of the p27 gene gained weight faster and grew larger than their normal littermates. Despite the increased size, the body was of comparable proportions to normal p27-containing mice, and no major abnormalities were evident. The increase in weight was not due to obesity, since all the organs weighed more or less proportionately more than that of comparable normal littermates. Thymus, spleen, and the pituitary were the most significantly enlarged, and the difference in the size of the thymus and spleen was evident by 4 weeks of age. The possibility of an endocrine contribution to the increased size of p27-null mice was also ruled out, since the mean serum levels of growth hormone (GH), insulin-like growth factor (IGF)-1, and IGF-1-binding protein were not significantly different from normal mice (KIYOKAWA et al. 1996). This suggests that the observed increase in size of organs is a result of increased numbers of cells that probably arose due to the absence of p27 to limit their proliferation.

3.3.2 Abnormalities of the Thymus in p27-Null Mice

The thymus of p27-null mice is grossly enlarged and sometimes covers the heart. In spite of the increased size of the thymus, the profiles of CD4/CD8 and $\alpha\beta$ T cell receptor are normal in the knockout mice. Furthermore, histopathological examination of the thymus reveals relatively normal architecture (NAKAYAMA et al. 1996), ruling out lymphoid malignancy as the reason for the increase in size. All populations of thymocytes are found to expand proportionately in the p27-null mice. One group detected an increase in the percentage of thymocytes in S phase of the cell cycle from a mean of 10% in normal mice to about 26% in p27-null mice (KIYOKAWA et al. 1996). No differences are apparent in the sensitivity of thymocytes from p27-null mice to spontaneous, γ-irradiation or dexamethasone-induced apoptosis. Taken together, these data suggest that the increased numbers of thymocytes in $p27^{-/-}$ mice are due to increased cell proliferation rather than decreased cell death (FERO et al. 1996).

Previous studies indicated that IL-2 treatment reduces the levels of p27 in peripheral T cells, suggesting that p27 affects the proliferation of T cells (NOURSE et al. 1994). When T cells from the lymph nodes of $p27^{-/-}$, $p27^{+/-}$, and $p27^{+/+}$ mice were tested for proliferation by anti-CD3 antibodies in the presence or absence of IL-2, all responded similarly. The absence of p27 does not augment the response of the $p27^{-/-}$ cells to mitogens, suggesting that either p27 exercises only a minor restraint in these cells or that another CDI is functioning in these cells in lieu of p27. Similarly, TGF-β treatment of T cells from the p27$^{-/-}$ mice results in an inhibition of proliferation, comparable in magnitude to the cells from normal mice (NAKAYAMA et al. 1996). The result is unexpected, since it had been fairly well established that TGF-β-mediated inhibition of cell proliferation is mediated mainly by p27. A possible interpretation of these results is that another CDI is functioning in place of p27, probably as a result of adaptation to the missing p27.

In spite of the grossly enlarged structure of the thymus, there are no striking abnormalities in the peripheral lymphoid organs. A few of the mice had approximately twofold more T cells in their spleen and lymph nodes, but the levels of B cells were unchanged (FERO et al. 1996). It is notable that the majority of the lymphocytes in the spleen and the lymph nodes remained in G_0 phase, suggesting that p27 was not necessary for maintaining these cells in the growth-arrested state (FERO et al. 1996). It is possible that the increase in the numbers of lymphocytes is a result of the increased number of cycles the precursors go through before exiting the cell cycle, since p27 is absent to apply the necessary constraint in growth.

Although the peripheral blood counts of erythrocytes, neutrophils, monocytes, and platelets of p27$^{-/-}$ mice are comparable to normal mice, there are considerably more hematopoietic progenitors of all types (FERO et al. 1996). There are increased amounts of granulocyte–macrophage, erythroid, and megakaryotic progenitors in both bone marrow and spleen. The increases are more pronounced in the spleen than in the marrow. These observations suggest an increase in splenic cellularity and may be indicative of a role for p27 in early stages of cell lineage determination.

3.3.3 Pituitary Hyperplasia in p27$^{-/-}$ Mice

Adenomas of the pituitary gland are observed in p27 null mice along with general hyperplasia (NAKAYAMA et al. 1996; KIYOKAWA et al. 1996; FERO et al. 1996). The pituitaries of the null mice weigh twice that of normal mice at 4 weeks of age. In one set of mice, the intermediary lobes of the pituitary were found to be highly hyperplastic, while the anterior and posterior lobes were relatively normal (NAKAYAMA et al. 1996). About half of the p27-null mice in one set of experiments developed pituitary adenomas originating from the intermediary lobes. There were a large number of atypical cells in the intermediary lobe that showed irregularly shaped nuclei and increased staining of the chromatin. Such atypical cells tended to form cysts with extensive hemorrhage, sometimes leading to the formation of hematoma. These features led to the diagnosis of these lesions as benign pituitary adenoma (NAKAYAMA et al. 1996). There were no signs of invasion or metastasis even at 7 months of age.

Closer examination of the adenoma by electron microscopy confirmed the finding that all the cells in the adenoma were derived from the intermediary lobe (FERO et al. 1996). The nuclei appeared to have more condensed chromatin than in normal cells. The abnormal tissue stained positively for α-melanocyte-stimulating hormone, β-endorphin, and adrenocorticotropic hormone, all products of the pars intermedia. No staining was observed for hormones produced by the anterior lobe of the pituitary in the adenomas (FERO et al. 1996; KIYOKAWA et al. 1996).

It is worth noting that mice heterozygous for the Rb gene also develop pituitary tumors (JACKS et al. 1992; E.Y.-H. LEE et al. 1992; WILLIAMS et al. 1994a, b). This could imply that Rb and p27 may be functioning in the same pathways of cell cycle regulation, and their regulation is mandatory in specific tissues such as the pituitary.

3.3.4 Reproductive System Abnormalities in p27$^{-/-}$ Mice

p27$^{-/-}$ mice have testes and ovaries twice the size of normal mice (NAKAYAMA et al. 1996). The testes have a normal architecture despite the considerable hyperplasia. The males are fertile and were able to impregnate control females, suggesting that normal development and function of spermatids occurs in the testes.

The female p27$^{-/-}$ mice had prolonged estrous cycles, were infertile, and could not carry pregnancy to term (NAKAYAMA et al. 1996; KIYOKAWA et al. 1996; FERO et al. 1996). This occurs despite the fact that these mice are capable of mating, as indicated by the appearance of vaginal plugs. Mice killed 3.5 days after the formation of the plug were found to have morula-stage embryos; these embryos developed normally when transferred to pseudopregnant normal mice. This result suggests that ovulation and fertilization do occur in p27$^{-/-}$ mice and that these processes do not require p27 (KIYOKAWA et al. 1996).

Defects in normal hormonal responses are suspected to cause the inability of the p27$^{-/-}$ mice to carry pregnancy to full term. Since corpus luteum formation plays a major role in secreting progesterone and other hormones required for normal pregnancy, its status was evaluated (KIYOKAWA et al. 1996). In normal mice, granulosa cells differentiate into progesterone-secreting luteal cells, and progesterone can be detected in the corpus luteum but not in the granulosa cells. In contrast, ovaries of p27$^{-/-}$ mice have defective formation of the corpus luteum, strongly suggesting that the transition of proliferating granulosa cells into differentiated and growth-arrested luteal cells requires p27 (KIYOKAWA et al. 1996).

3.3.5 Effects of p27 Knockout on the Retinal Structure

The retinas of p27$^{-/-}$ mice display a significant disorganization of the cellular layer pattern in the neural retina (NAKAYAMA et al. 1996). The outer granular layer, which consists of nuclei of photoreceptor cells, invaded the layer of rods and cones beyond the outer limiting membrane. To elucidate the cause of this highly focused abnormality, the expression pattern of p27 in the retina was examined immunohistochemically. Specific expression of the p27 was detected in the photoreceptor cells of normal mice. This lends support to the notion that the absence of p27 leads to the disorganization of the cellular layer of the retina (NAKAYAMA et al. 1996).

The finding that p27$^{-/-}$ mice have defective retinas is striking in the context of normal Rb function, since Rb inactivation leads to retinoblastomas reminiscent of the type seen in p27$^{-/-}$ mice (JACKS et al. 1992; E.Y.-H. LEE et al. 1992). This supports the hypothesis that, at least in some tissues, p27 and Rb regulate cell proliferation via the same pathway.

3.3.6 Growth Arrest Is Normal in p27$^{-/-}$ Fibroblasts

Mouse embryo fibroblasts prepared from normal or p27$^{-/-}$ mice grow normally and at comparable rates (NAKAYAMA et al. 1996). G$_1$-phase arrest occurs in both cell types following contact inhibition, serum deprivation, γ-irradiation, and chemical inhibition of the cell cycle. These observations, although surprising, suggest that growth arrest by these treatments does not require p27 or that, in these cells, another CDK inhibitor is performing the inhibitory function normally performed by p27.

3.3.7 Mechanism of Cyclin-Dependent Kinase Inhibition by p27

It is well established that p21 and p27 bind to cyclin–CDK complexes and inhibit their activity, unlike the INK4 family inhibitors, which prevent CDK from binding to cyclins. Resolution of the crystal structure of the amino-terminal 69 amino acids of p27 bound to the cyclin A–p27 complex has provided important mechanistic insights into the physical basis of p27 action (RUSSO et al. 1996). It is apparent that the peptide is draped across the top of the two subunits in an extended conformation. The furthest amino-terminal region that contains a conserved LFG motif interacts with a binding pocket of cyclin A (RUSSO et al. 1996; MORGAN 1996). The carboxy-terminal region of the p27 peptide disrupts the configuration of the cdk2 kinase and binds to the catalytic cleft, preventing the binding of ATP. It appears that the binding of p27 to cyclin–cdk2 completely alters the configuration of the kinase and ensures the complete inhibition of function by multiple interactions (RUSSO et al. 1996).

4 Cyclin-Dependent Kinase Inhibitor p57KIP2

p57KIP2 is a structurally distinct member of the p21/p27 family cloned using two different approaches (M.H.J. LEE et al. 1995; MATSUOKA et al. 1995). Low-stringency screening of a mouse cDNA library using a p21 probe yielded a mouse cDNA that was distinct from p21 and p27 (M.H.J. LEE et al. 1995). A modified yeast two-hybrid system was also used to identify mouse proteins that interact with cyclin D$_1$, resulting in the isolation of the p57KIP2 clone (MATSUOKA et al. 1995). This clone was then used to screen a human cDNA library to obtain the corresponding human gene, which has structural features different from the mouse protein.

The mouse p57 protein has a predicted length of 348 amino acids and has four distinct amino acid sequence domains (M.H.J. LEE et al. 1995). The amino-terminal domain has the maximum amount of sequence homology to p21 and p27 in a region that has been shown to be necessary and sufficient to inhibit CDK activity. This is followed by a proline-rich region that has a series of alternating prolines.

The third domain is highly acidic, and every fourth residue is glutamic or aspartic acid. A carboxy-terminal domain is also evident that has homology to the C terminus of p27 and contains a QT box (M.H.J. LEE et al. 1995).

The human p57KIP2 homologue is 316 amino acids long and shares the amino- and carboxy-terminal domains with the mouse protein (MATSUOKA et al. 1995). In the place of the two internal domains, there is an alternating repeat of proline–alanine residues, termed the PAPA repeat, the function of which is unknown. A differentially spliced form of human p57 is also observed that has amino acid residues 1-96 corresponding to the CDK-binding domain (MATSUOKA et al. 1995). It is possible that different spliced forms of human and mouse p57 messages exist, but it is not clear whether they differ in function. p57 is a nuclear protein and contains a putative nuclear localization signal toward the C-terminal region of the protein.

p57KIP2 can bind to cyclin–CDK complexes in a cyclin-dependent manner (MATSUOKA et al. 1995). It can efficiently bind cdk2, cdk3, and cdk4, but binds weakly to the cdk6–cyclin D_2 complex. There is no detectable binding to cyclin H–cdk7 (MATSUOKA et al. 1995). p57 inhibits the kinase activity of cyclin E–cdk2, cyclin A–cdk2, cyclin E–cdk3, and cyclin D_2-cdk4 complexes (M.H.J. LEE et al. 1995; MATSUOKA et al. 1995). The inhibitory effects on cyclin B–cdc2 and cyclin D_2–cdk6 complexes are marginal. Consistent with the inhibition of CDK activities, overexpression of p57KIP2 leads to an arrest of the cell cycle in the G_1 phase (M.H.J. LEE et al. 1995; MATSUOKA et al. 1995). Unlike p21, there is no induction of p57 by the p53 protein (MATSUOKA et al. 1995).

4.1 Differentiation and Development

p57 displays tissue-specific expression patterns and is not ubiquitously expressed like p21 and p27 (M.H.J. LEE et al. 1995; MATSUOKA et al. 1995). This gene is expressed in heart, brain, lung, skeletal muscle, kidney, pancreas, and testis, as determined by northern blot analysis of human RNA. The maximum level of expression is in the placenta, with low to undetectable RNA levels in liver and spleen. The major transcript is approximately 1.5–1.7 kb long, but a 6- to 7-kb transcript is also apparent in certain tissues. The relevance of this larger transcript is not known, since it cannot be detected when a probe for the CDK-binding domain is used (M.H.J. LEE et al. 1995).

The in vivo expression of p57 message has been examined during mouse embryogenesis (MATSUOKA et al. 1995). p57 is expressed in brain, lens epithelium, Rathke's pouch, and the otocyst. Expression is also abundant in the dermamyotome and later in skeletal muscles and cartilage of the cervical region and forelimbs. The expression pattern of p57 appears to loosely correspond with that of p21, although in tissues such as brain, epidermis, hair follicles, and eyes there is no overlapping expression. These results indicate a good correlation between p57 expression and the differentiated state of cells during development (MATSUOKA et al. 1995). It is reasonable to conclude that p57 does indeed play a role in the exit of cells from the cell cycle and contributes to the process of terminal differentiation during mouse development.

5 INK4 Family of Cyclin-Dependent Kinase Inhibitors

The first member of the INK4 family of CDI was cloned using a yeast two-hybrid system with cdk4 as the bait and turned out to be a gene coding for an 148-amino acid polypeptide referred to as p16 (SERRANO et al. 1993). The most striking feature of this protein is the presence of four tandemly repeated ankyrin repeats, each approximately 32 amino acids in length. p16 has been shown to be a strong inhibitor of cyclin D–cdk4 kinase and bound the respective CDK directly (SERRANO et al. 1993). There is no binding to cdk2 when examined by co-immunoprecipitation. This finding was confirmed by performing a yeast two-hybrid analysis, which indicates that p16 can bind efficiently to cdk4, but cannot bind to cdk2, cdk5, cdc2, or PCNA. Further experimentation demonstrated that p16 can bind to cdk6, thereby inhibiting its activity. Three additional members of this family have been cloned and characterized (p15INK4B, p18INK4C, and p19INK4D), all possessing the characteristic ankyrin repeats (HANNON and BEACH 1994; K. GUAN et al. 1994; CHAN et al. 1995; HIRAI et al. 1995).

p16INK4A and p15INK4B are tandemly linked on chromosome 9p21 (HANNON and BEACH 1994), a region that sustains loss of heterozygosity and homozygous deletions in multiple human tumors (KAMB et al. 1994; NOBORI et al. 1994; OKOMOTO et al. 1994). The p16 locus gives rise to two distinct transcripts from two different promoters. Both transcripts possess a distinct 5'-exon, E1a or E1b, which is spliced on to two common exons, E2 and E3 (DURO et al. 1995; STONE et al. 1995; QUELLE et al. 1995a, b). The E1a-containing transcript encodes p16INK4A, and the E1B-containing transcript encodes p19ARF (alternate reading frame). p19ARF appears to arrest cell cycle both at G_1/S and at G_2/M phase, probably through mechanisms distinct from cdk4 inhibition (QUELLE et al. 1995).

p15 expression is induced about 30-fold by TGF-β in human keratinocytes, suggesting that it may play a role in TGF-β-mediated growth arrest (HANNON and BEACH 1994). p15 appears to be as specific as p16 in its ability to target cyclin D–cdk4/cdk6 kinase activities. Thus it appears that different INK4 proteins may function to induce cell cycle arrest, but may do so under different physiologic conditions in response to different signals. p18INK4C and p19INK4D are located at different loci, 1p32 and 19p13, respectively (GUAN et al. 1994; CHAN et al. 1995).

p16 is mutated in the majority of cultured human tumor cell lines, but the role of p16 inactivation in the genesis of human cancer has been questioned. It has now been confirmed that p16 is inactivated by deletions, rearrangements, or point mutations in many primary tumor types, especially acute lymphocytic leukemia (ALL) and melanomas (DURO et al. 1996; OKUDA et al. 1995b; HEBERT et al. 1994). Inactivation of p16 by chromosome translocations has been observed in certain cases (DURO et al. 1996), and there is evidence suggesting that p16 is inactivated by methylation of a CpG island spanning exon 1 and part of exon 2 in colon cancer (GONZALEZ-ZULUETA et al. 1995).

Early studies indicate that the p16 gene is overexpressed in tumor cell lines that lack a functional Rb gene (SERRANO et al. 1993; PARRY et al. 1995; UEKI et al. 1996). Since p16 can inhibit only cdk4 and cdk6, and its only known substrate is the Rb protein, it is proposed that the major function of p16 in the cell is to maintain Rb in a functional state. This could explain the abundance of p16 in cells lacking Rb; since there is no Rb to be inactivated, there might not be any need for cdk4/cdk6 activity to be present in the cell, and the excess p16 functions to maintain cdk4 in an inactive state (for a review, see in SHERR and ROBERTS 1995). Support for this idea comes from the observation that p16-mediated inhibition of cell cycle progression is completely dependent on the presence of a functional Rb protein (LUKAS et al. 1995). This was demonstrated by microinjection of a wild-type or mutant p16 into human fibroblasts, resulting in an enrichment of cells in G_1 phase in the presence of wild-type p16. In contrast, transfection of p16 has no effect on Saos-2 cells that lack a functional Rb gene, but there is a significant increase in G_1-phase cells when transfected into SKLMS-1 and U2OS cells (LUKAS et al. 1995; TAM et al. 1995). Similar results have been obtained using mouse cell lines, in which p16 can inhibit 3T3 cells or mouse embryo fibroblasts from dividing, but cannot induce a G_1-phase arrest of fibroblasts derived from $RB^{-/-}$ mice.

Consistent with the idea that p16 contributes primarily to maintenance of Rb in a functional state, the p16 and Rb genes are mutated in different subsets of tumors, and their inactivation is totally mutually exclusive. Since the main function of p16 appears to be to maintain Rb in a functional state, it can be envisioned that there is no need for mutating two genes that function in the same pathway to control cell cycle. Alternatively, mutating the p16 gene will lead to an inactivation of Rb by cdk4/cdk6, which would be equivalent to a mutation of the RB gene. Similarly, if the Rb gene is mutated in a cell, the presence of functional cdk4 would be superfluous, and hence an inactivation of p16 would not be necessary.

The role of p16 in tumor suppression became more apparent when it was shown that tumor-derived mutant p16 alleles are defective in cell cycle inhibition (KOH et al. 1995). Out of nine different alleles derived from tumors, four were completely impaired in their ability to induce cell cycle arrest; two were similar to the wild-type allele, and three had intermediate activities. Two of the mutants with the intermediate levels of growth suppression were completely effective in inhibiting cdk4 kinase activity; the third had no apparent effect on cdk3 activity, but could bring about intermediate levels of growth suppression (KOH et al. 1995). These observations might indicate that, although the ability of p16 to inhibit cdk4 activity is primarily responsible for effecting growth suppression, additional steps or pathways are also necessary for p16 to exert complete inhibition of cell proliferation.

p16 has a profound inhibitory effect on Ras-induced proliferation and transformation of REF-52 cells. Microinjection of a plasmid encoding activated Ha-ras (V12 Ras) induces S-phase entry and DNA synthesis effectively in REF-52 cells, but this response can be effectively blocked by a p16 expression vector (SERRANO et al. 1995). An antisense p16 vector cannot block the S-phase entry. The specificity of the inhibition of V12 Ras-induced proliferation by p16 was tested by including a catalytically inactive cdk4, cdk4-K35M. This mutant cdk4 binds p16 efficiently, but cannot phosphorylate Rb. Microinjection of this mutant along with p16 significantly relieves p16-induced repression of cell proliferation, suggesting that p16 functions mainly by targeting cyclin D–cdk4 kinase (SERRANO et al. 1995). An interesting observation in this study is that p16 can effectively block transformation by c-myc plus V12 Ras, but has no effect on transformation induced by adenovirus E1A plus V12 Ras. These results provide further evidence that Rb is the major functional target of p16, and once Rb is inactivated by E1A, p16 has no effect as a growth suppressor (SERRANO et al. 1995). Thus multiple lines of evidence place p16 as a growth suppressor that functions upstream of Rb in the same pathway.

5.1 p16 and Cellular Differentiation

CDK and CDI play important roles in myogenic differentiation. Studies carried out in Andrew Lassar's laboratory indicate that cyclin D–CDK complexes can phosphorylate and inactivate MyoD (SKAPEK et al. 1995). Although MyoD is present in myoblasts at all stages of the cell cycle, its differentiation-inducing effects are restricted to G_1 phase of the cell cycle. Transient transfection of cyclin D_1, but not cyclins A, B_1, B_2, or E, inhibits the ability of MyoD to induce muscle-specific genes as well as a myogenic phenotype. Ectopic expression of p16 or p21 can reverse cyclin D-mediated inhibition of MyoD function as well as differentiation (SKAPEK et al. 1995). These results suggest that cyclin D_1 inhibits MyoD function by a CDK-dependent mechanism and that inhibitors of this kinase may play a role in facilitating the process of differentiation. However, it is worth noting that earlier studies indicate a definitive role for Rb and Rb family members in myogenic differentiation; it is therefore possible that the effects of cyclin D and p16 occur via the Rb pathway (CARUSO et al. 1993; Y. GU et al. 1993; SCHNEIDER et al. 1995).

5.2 Role of p16 in Development

To define the role of p16 during murine development, mice were engineered with a targeted deletion of the p16 gene (SERRANO et al. 1996). These mice lack both p16 and the p19ARF products, since exons 2 and 3 are deleted. Surprisingly, the homozygous null mice are viable and fertile, and there are no gross phenotypic changes. The p16-null mice have a slightly lighter coat color, but it is not clear whether the p16 gene contributes to this phenotype. Organs from $p16^{-/-}$ mice are normal, although there appears to be a slight proliferative expansion of the white pulp of the spleen, and there are megakaryocytes and lymphoblasts in the red pulp (SERRANO et al. 1996). These features are consistent with abnormal extramedullary hematopoiesis in the spleen of young mice. It is possible that p16 can negatively regulate the proliferation of some hematopoietic progenitors or their lineages. Other than this, there appears to be no abnormalities in the development of the p16 null mice.

5.3 Tumor Susceptibility of p16-Null Mice

The role of p16 in tumor suppression was ascertained by exposing heterozygous intercrosses to different carcinogens and comparing their susceptibility to that of wild-type, heterozygote, and null mice (SERRANO et al. 1996). Over 500 mice, ranging in age from 1 to 10 months and representing all three genotypes, were monitored for tumor development after exposure to 9,10-dimethyl-1,2-benzanthracene (DMBA), irradiation with UVB, or a combination of both treatments. Ninety percent of $p16^{-/-}$ mice exposed to the combined carcinogen treatment showed obvious tumors or compromised health. In comparison, only 13% of the heterozygotes and none of the wild-type littermates exhibited signs of tumors or disease. The average latency for confirmed tumors in the $p16^{-/-}$ mice was 9 weeks, whereas the $p16^{+/-}$ mice showed a latency of about 15 weeks. In the group of animals treated with DMBA alone, the survival rate was better than in those exposed to DMBA and UVB; those exposed to UVB alone behaved similarly to wild-type mice (SERRANO et al. 1996).

Although the $p16^{-/-}$ mice develop a wide spectrum of tumors, they are dominated by two categories of tumors. One type develops in the subcutis and invades the underlying musculature; it consists of anaplastic spindle cells and is probably a form of fibrosarcomas. The second category of tumors exhibits a more general distribution and is lymphoid in nature. The architecture of the affected lymph nodes is effaced by small, round cells with hyperchromatic nuclei and displays positive staining with CD45 and B220 markers. Since these cells do not react with anti-CD3, the tumors appear to be B cell lymphomas. The spectrum of tumors in $p16^{-/-}$ mice exposed to carcinogens is same as those in mice not exposed to carcinogens. Carcinogen exposure does not alter the spectrum of tumors, but renders the mice more susceptible to carcinogenesis.

5.4 Properties of $p16^{-/-}$ Mouse Embryo Fibroblasts

Cultures of mouse embryo fibroblasts from $p16^{-/-}$ mice are indistinguishable from $p16^{+/+}$ mouse embryo fibroblasts, except that they grow faster and achieve higher saturation densities (SERRANO et al. 1996). Analyses of the different cell cycle stages of the $p16^{-/-}$ fibroblasts indicate a moderate increase in the S-phase population accompanied by a corresponding decrease in the G_0/G_1 cells. It is also apparent that $p16^{-/-}$ early-passage fibroblasts form colonies more efficiently that wild-type ones, suggesting that $p16^{-/-}$ cells can overcome senescence more easily than wild-type mouse embryo fibroblasts. This conclusion is supported by the observation that, unlike normal mouse embryo fibroblasts, $p16^{-/-}$ cultures maintained a rapid and constant growth rate, and a slow-growing senescent stage was not reached even after 25 passages.

Studies were performed to evaluate the susceptibility of p16$^{-/-}$ fibroblasts to oncogenic transformation. Normal primary cell lines require two separate oncogenes to cooperate and induce transformation. Transfection of Ha-Ras or myc into early-passage p16$^{+/+}$ or $^{+/-}$ cells does not produce any morphologically transformed foci. Surprisingly, introduction of Ha-Ras alone into the p16$^{-/-}$ cells induces the formation of foci. In contrast, the c-myc gene is unable to induce foci formation by itself in p16$^{-/-}$ fibroblasts. Further evidence that a single oncogene can induce foci formation was provided using activated Raf-1 (Raf-1 BXB), which is as efficient as Ha-Ras in foci formation. Raf-1 BXB cannot generate foci formation in normal mouse embryo fibroblasts. These findings suggest that p16 inactivation can cooperate with activated Ha-Ras to induce transformation and that loss of p16 may represent an immortalization event. On the whole, the studies using p16-null mice confirm the hypothesis that p16 is a major regulator of cell proliferation and that it can function as a true tumor suppressor gene (SERRANO et al. 1996).

6 p18INK4C

p18INK4C was first cloned using a yeast two-hybrid system to identify proteins that interact with cdk6 (K. GUAN et al. 1994). A gene corresponding to an 18 116-kDa protein was isolated using this strategy and displays 38% sequence homology to p16 over a 152-amino acid region and 42% identity to p15 over a 129-amino acid span. The p18 gene has homology to the human Notch family of genes, especially TAN-1, in the ankyrin repeat region. It is not clear whether the ankyrin repeat of p18 is functionally distinct in any fashion from those of other INK4 proteins. Like p16, p18 can associate with cdk4 and cdk6 in vitro and in vivo and can inhibit their kinase activity (HIRAI et al. 1995). Northern blot analysis indicates five distinct p18 transcripts, whose tissue distribution varies considerably (K. GUAN et al. 1994). The significance of this observation is presently unclear. p18 induces a G_1-phase arrest similar to p16, and its growth-inhibitory effects depend on the presence of a functional Rb gene. The role of p18 in differentiation and development remains to be elucidated.

7 p19INK4D

An orphan steroid receptor Nur 77 has been identified that induces T cell receptor-mediated G_1-phase arrest and apoptosis (CHAN et al. 1995). Attempts to clone proteins that interact with Nur 77 using a yeast two-hybrid screen lead to the cloning of a gene with considerable similarity to p16INK4A, termed p19INK4D. The human and mouse p19 genes are highly homologous and show 81% identity (CHAN et al. 1995). The deduced sequence of human p19 has 48% identity with p16 over a stretch of 130 amino acids; sequence comparisons with other INK4 proteins strongly suggest that p19 is indeed part of the INK gene family. Its association with the Nur77 protein and the functional consequence of such an interaction is still a mystery (CHAN et al. 1995).

p19 can associate with cdk4 in vitro, but not with cdk2, cdc2, or cyclins A, B, D, and E. It can inhibit cyclin D–cdk4 activity, but not cyclin E–cdk2, and can associate with cdk4 and cdk6 in vivo. Ectopic expression of p19 can induce a G_1-phase arrest (HIRAI et al. 1995). The dependence of p19 on Rb protein for inducing growth arrest is not clear. Cell cycle analysis of p19 expression in cultured macrophages and fibroblasts indicates that the p19 protein is induced during S phase. Since p16 is not significantly modulated during the cell cycle and cannot be detected in many mouse tissues, it has been suggested that p19, rather than p16, may function to regulate the cell cycle in specific mouse tissues (HIRAI et al. 1995; CHAN et al. 1995). The role of p19 in cellular differentiation and development is currently unknown.

8 Future Outlook

The identification of CDI provides a mechanistic framework for examining how CDK are functionally inactivated without cyclin degradation. The activation of CDI occurs when a cell is induced to stop its normal cycling and growth arrest, such as p21 induction as a function of DNA damage and differentiation, p27 induction following cell–cell contact, and elevated p15 expression in T cells exposed to TGF-β. The absence of these important negative cell cycle regulators can be envisioned to induce numerous adverse consequences for a cell, including increased potential for generating mutations and fostering genomic instability, uncontrolled cell division leading to cancer progression, possibly culminating in metastasis, and lack of growth control resulting in altered programs of differentiation.

The development of eukaryotic organisms is contingent upon the appropriate temporal formation of functionally differentiated tissues. For most tissue types, the molecular and biochemical mediators of cell lineage determination that are responsible for terminal differentiation and development are not known. In the case of specific differentiation programs, such as skeletal myogenesis and adipogenesis, genes that can function as direct inducers of the differentiated phenotype have been identified, cloned, and characterized. A property shared by all terminally differentiated cells is an inability to reenter the cell cycle, even in the presence of mitogenic stimuli. Regulation of cell cycle progression is controlled at multiple levels, including the quantity and status of phosphorylation and dephosphorylation of

CDK and the concentration of specific CDI. In this review, we have attempted to survey those studies leading to the conclusion that CDI are important molecules in the differentiation process of specific target cells and in development. What is most apparent from the various studies, particularly those involving null phenotype mice, is that the various CDI may have redundant and overlapping properties in regulating and maintaining cell differentiation and development.

Specific CDI can directly suppress the growth of cancer cells. In this context, these molecules may prove useful for the targeted therapy of specific malignancies. This strategy has been used with some preliminary success by incorporating CDI, such as p21, in adenovirus vectors. Infection of human cancer cells, including breast and lung carcinoma and melanoma cells, with p21-expressing replication-defective adenoviruses (Ad.p21) results in the loss of tumorigenic potential. Moreover, injection of Ad.p21 directly into tumors or peritumorally can induce a suppression in established human tumor xenografts in nude mice. By using specifically engineered CDI and tissue-specific expression vectors, it may also be possible in the future to specifically target CDI for defined therapeutic applications. In addition, CDI may prove useful as markers for the induction of a chemotherapeutic response in cancer cells and as diagnostic indicators for induction of specific differentiation programs.

An understanding of cell cycle on a molecular and biochemical level will provide new opportunities for comprehending cancer and may identify important new target molecules for preventing tumor development and progression. The combination of this information and the continued improvement in gene-based therapies and small-molecule approaches for cancer therapy should result in valuable methodologies for improving the negative prognosis associated with many human malignancies.

Acknowledgements. Support for specific studies described in this review was provided in part by NCI-NIH grants CA63136 (to S.P.C.), CA35675 (to P.B.F.), and CA60999 (to A.G.), the Council for Tobacco Research (to A.G.), the Irma-Hirschl Trust (to S.P.C.), the Chernow Endowment (to P.B.F.), and the Samuel Waxman Cancer Foundation to (P.B.F.). This research was also assisted by the Sbarro Institute for Cancer Research and Molecular Medicine (A.G.). P.B.F. is a Chernow Research Scientist in the Departments of Pathology and Urology. Our appreciation is extended to the numerous colleagues that have contributed to the studies presented in this review, including Dr. Neil I. Goldstein, Dr. Steven Grant, Dr. Hongping Jiang, Dr. Jian Lin, Dr. Jiao Jiao Lin, and Dr. Zao-zhong Su. We thank Dr. Timothy McLachlan for assistance with the cell cycle figure.

References

Andres V, Walsh K (1996) Myogenin expression, cell cycle withdrawal and phenotypic differentiation are temporally separable events that precede cell fusion upon myogenesis. J Cell Biol 132:657–666
Aprelikova O, Xiong Y, Liu ET (1995) Both p16 and p21 families of cyclin-dependent kinase (CDK) inhibitors block the phosphorylation of cyclin-dependent kinases by the CDK-activating kinase. J Biol Chem 1270:18195–18197
Bates S, Peters G (1995) Cyclin D1 as a cellular proto-oncogene. Semin Cancer Biol 6:73–82
Benezra R, Davis RL, Lockshon D, Turner DL, Weintraub H (1990) The protein Id: a negative regulator of helix-loop-helix DNA binding proteins. Cell 61:49–59

Biggs JR, Kraft AS (1995) Inhibitors of cyclin-dependent kinase and cancer. J Mol Med 75:509–514
Bishop JM (1987) The molecular genetics of cancer. Science 235:305–311
Bishop JM (1991) Molecular themes in oncogenesis. Cell 64:235–248
Breitman T, Delonick D, Collins S (1980) Induction of differentiation of the human promyelocytic leukemia cell line (HL-60) by retinoic acid. Proc Natl Acad Sci USA 77:2936–2940
Brugarolas J, Chandrasekaran C, Gordon JI, Beach D, Jacks T, Hannon GJ (1995) Radiation-induced cell cycle arrest compromised by p21 deficiency. Nature 377:552–557
Bullrich F, MacLachlan TK, Sang N, Druck T, Veronese ML, Allen SL, Chiorazzi N, Koff A, Huebner K, Croce CM, Giordano A (1995) Chromosomal mapping of members of the cdc2 family of protein kinases, cdk3, cdk6, PISSLRE and PITALRE and a cdk inhibitor, $p27^{Kip1}$, to regions involved in human cancer. Cancer Res 55:1199–1205
Calautti E, Missero C, Stein PL, Ezzell RM, Dotto GP (1995) Fyn tyrosine kinase is involved in keratinocyte differentiation control. Genes Dev 9:2279–2291
Caruso M, Martelli F, Giordano A, Felsani A (1993) Regulation of MyoD gene transcription and protein function by the transforming domains of the adenovirus E1A oncoproteins. Oncogene 8:267–278
Castresana JS, Rubi MP, Vazquez J, Idoate M, Sober AJ, Seizinger BR, Barnhill RL (1993) Lack of allelic deletion and point mutation as mechanisms of p53 activation in human malignant melanoma. Intl J Oncol 55:562–565
Chan FK, Zhang J, Cheng L, Shapiro DN, Winoto A (1995) Identification of human and mouse p19, a novel CDK4 and CDK6 inhibitor with homology to p16ink4. Mol Cell Biol 15:2682–2688
Chellappan SP (1994) The E2F transcription factor: role in cell cycle regulation and differentiation. Mol Cell Diff 2(3):201–220
Chen J, Jackson PK, Kireschner MW, Dutta A (1995) Separate domains of p21 involved in the inhibition of Cdk kinase and PCNA. Nature 374:386–388
Chen X, Bargonetti J, Prives C (1995) p53, through p21 (WAF1/CIP1), induces cyclin D1 synthesis. Cancer Res 55:4257–4263
Clark W (1991) Tumor progression and the nature of cancer. Br J Cancer 64:631–644
Clurman BE, Bruns G (1995) Cell cycle and cancer. J Natl Cancer Inst 87:1499–1501
Coats S, Flanagan WM, Nourse J, Roberts JM (1996) Requirement of p27Kip1 for restriction point control of the fibroblast cell cycle. Science 272:877–880
Cobrinik D, Dowdy SF, Hinds PW, Mittnacht S, Weinberg RA (1992) The retinoblastoma protein and the regulation of cell cycling. Trends Biochem Sci 17:312–315
Collins SJ (1987) The HL-60 promyelocytic leukemia cell line: proliferation, differentiation, and cellular oncogene expression. Blood 70:1233–1244
Collins SJ, Ruscetti FW, Gallagher R, Gallo R (1978) Terminal differentiation of the human promyelocytic leukemia cells induced by dimethylsulfoxide and other polar compounds. Proc Natl Acad Sci USA 75:2458–2462
Coppola JA, Lewis BA, Cole MD (1990) Increased retinoblastoma gene expression is associated with late stages of differentiation in many different cell types. Oncogene 5:1731–1733
De Bondt HL, Rosenblatt J, Jancarik J, Jones HD, Morgan DO, Kim SH (1993) Crystal structure of cyclin-dependent kinase 2. Nature 363:595–602
Del Sal G, Murphy M, Ruaro EM, Lazarevic D, Levine AJ, Schneider C (1996) Cyclin D1 and p21/waf1 are both involved in p53 growth suppression. Oncogene 12:177–185
Desai D, Wessling HC, Fisher RP, Morgan DO (1995) Effects of phosphorylation by CAK on cyclin binding by CDC2 and CDK2. Mol Cell Biol 15:345–350
Deng C, Zhang P, Harper JW, Elledge SJ, Leder P (1995) Mice lacking $p21^{CIP1/WAF1}$ undergo normal development, but are defective in G1 checkpoint control. Cell 82:675–684
DiGiuseppe JA, Redstone MS, Yeo CJ, Kern SE, Hruban RH (1995) p53-independent expression of the cyclin-dependent kinase inhibitor p21 in pacreatic carcinoma. Am J Pathol 147:884–888
Donehower LA, Harvey M, Slagle BL, McArthur MJ, Montgomery CA Jr, Butel JS, Bradley A (1992) Mice deficient for p53 are developmentally normal but susceptible to spontaneous tumors. Nature 356:215–221
Dover R, Wright NA (1991) The cell proliferation kinetics of the epidermis. In: Goldsmith LA (ed) Physiology, biochemistry, and molecular biology of the skin. Oxford University Press, New York, pp 239–265
Dracopoli NC, Sam B, Fountain JW, Schartl M (1992) A homolog of the Xiphophorus melanoma-inducing locus Xmrk maps to 6q27 a region frequently involved in rearrangements and deletions in human melanomas. Abstr Am J Hum Genet 54:A51

Dulic V, Kaufmann WK, Wilson SJ, Tlsty TD, Lees E, Harper JW, Elledge SJ, Reed SI (1994) p53-dependent inhibition of cyclin-dependent kinase activities in human fibroblasts during radiation-induced G1 arrest. Cell 76:1013–1023

Duro D, Bernard O, Della Valle V, Berger R, Larsen CJ (1995) A new type of p16INK4/MTS1 gene transcript expressed in B-cell malignancies. Oncogene 11:21–29

Duro D, Bernard O, Della Valle V, Leblanc T, Berger R, Larsen CJ (1996) Inactivation of the P16INK4/MTS1 gene by a chromosome translocation t(9;14)(p21–22;q11) in an acute lymphoblastic leukemia of B-cell type. Cancer Res 56:848–854

Elbendary A, Berchuck A, Davis P, Havrilesky L, Bast RC, Iglehart JD, Marks JR (1994) Transforming growth factor ß1 can induce CIP1/WAF1 expression independent of the p53 pathway in ovarian cancer cells. Cell Growth Diff 5:1301–1307

El-Deiry WS, Tokino T, Velculesco VE, Levy DB, Parsons R, Trent JM, Lin D, Mercer WE, Kinzler KW, Vogelstein B (1993) WAF1, a potential mediator of p53 tumor suppression. Cell 75:817–825

Elledge SJ, Harper JW (1994) Cdk inhibitors: on the threshold of checkpoints and development. Curr Opin Cell Biol 6:847–852

Ewen ME (1994) The cell cycle and the retinoblastoma protein family. Cancer Metastas Reviews 13:45–66

Ewen ME, Sluss HK, Sherr CJ, Matsushime H, Kato J, Livingston DM (1993) Functional interactions of the retinoblastoma protein with mammalian D-type cyclins. Cell 73:487–497

Fero ML, Rivkin M, Tasch M, Porter P, Carow CE, Firpo E, Polyak K, Tsai LH, Broudy V, Perlmutter RM, Kaushansky K, Roberts JM (1996) A syndrome of multiorgan hyperplasia with features of gigantism, tumorigenesis, and female sterility in p27(Kip1)-deficient mice. Cell 85:733–744

Filvaroff E, Stern DF, Dotto GP (1990) Tyrosine phosphorylation is an early and specific event involved in primary keratinocyte differentiation. Mol Cell Biol 10:1164–1173

Firpo EJ, Koff A, Solomon MJ, Roberts JM (1994) Inactivation of cdk2 inhibitor during interleukin 2-induced proliferation of human T-lymphocytes. Mol Cell Biol 14:4889–4901

Fisher PB, Schachter D, Abbott RE, Callaham MF, Huberman E (1984) Membrane lipid dynamics in human promyelocytic leukemia cells sensitive and resistant to 12-O-tetradecanoyl-phorbol-13-acetate induction of differentiation. Cancer Res 44:5550–5554

Fisher PB, Prignoli DR, Hermo H Jr, Weinstein IB, Pestka S (1985) Effects of combined treatment with interferon and mezerein on melanogenesis and growth in human melanoma cells. J Interferon Res 5:11–22

Fisher RP, Morgan DO (1994) A novel cyclin associates MO15/CDK7 to form the CDK-activating kinase. Cell 78:713–724

Flores-Rozas H, Kelman Z, Dean FB, Pan Z, Harper WJ, Elledge SJ, O'Donnell M, Hurwitz J (1994) Cdk-interacting protein 1 directly binds with proliferating cell nuclear antigen and inhibits DNA replication catalyzed by the DNA polymerase and holoenzyme. Proc Natl Acad Sci USA 91:8655–8659

Fountain JW, Bale SJ, Housman DE, Dracopoli NC (1990) Genetics of melanoma. Cancer Surv 9:645–671

Freemerman AJ, Vrana JA, Tombes RM, Jiang H, Fisher PB, Chellappan SP, Grant S (1997) Effects of antisense p21 (WAF1/CIP1/MDA-6) expression on the induction of differentiation and drug-mediated apoptosis in human myeloid leukemia cells (HL-60). Leukemia 11:504–513

Galktionov K, Jesus C, Beach D (1995) Raf-1 interaction with Cdc25 phosphatase ties mitogenic signaling to cell cycle activation. Genes Dev 9:1046–1058

Gautier J, Solomon M, Booher R, Bazan J, Kirshner M (1991) Cdc25 is a specific tyrosine phosphatase that directly activates p34cdc2. Cell 67:197–211

Giordano A, Kaiser H (1996) The retinoblastoma gene: its role in cell cycle and cancer. In Vivo, 10:223–228

Goldstein S (1990) Replicative senescence: the human fibroblast comes of age. Science 249:1129–1133

Gonzalez-Zulueta M, Bender CM, Yang AS, Nguyen T, Beart RW, Van Tornout JM (1995) Methylation of the 5' CpG island of the p16/CDKN2 tumor suppressor gene in normal and transformed human tissues correlates with gene silencing. Cancer Res 55:4531–4535

Grana X, Reddy EP (1995) Cell cycle control in mammalian cells: role of cyclins, cyclin dependent kinases (CDKs), growth suppressor genes and cyclin-dependent kinase inhibitors (CKIs). Oncogene 11:211–219

Greenblatt MS, Bennett WP, Hollstein M, Harris CC (1994) Mutations in the p53 tumour suppressor gene: clues to cancer etiology and molecular pathogenesis. Cancer Res 54:4855–4878

Gu W, Schneider JW, Condorelli G, Nadal-Ginard B (1993) Interaction of myogenic factors and the retinoblastoma protein mediates muscle cell commitment and differentiation. Cell 72:309–324

Gu Y, Rosenblatt J, Morgan DO (1992) Cell cycle regulation of CDK2 activity by phosphorylation of Thr160 and Tyr15. EMBO J 11:3995–4005

Gu Y, Turck CW, Morgan DO (1993) Inhibition of CDK2 activity in vivo by an associated 20 K regulatory subunit. Nature 366:707–710

Guan K, Jenkins CW, Li Y, Nichols MA, Wu X, O'Keefe CL, Matera AG, Xiong Y (1994) Growth suppression by p18, a p16INK/MTS1 and p14INK4/MTS2-related CDK6 inhibitor, correlates with wild-type RB function. Genes Dev 8:2939–2952

Guan X, Meltzer PS, Cao J, Trent, JM (1992) Rapid generation of region-specific genomic clones by chromosome microdissection: isolation of DNA from a region frequently deleted in malignant melanoma. Genomics 14:680–684

Guo K, Wang J, Andres V, Smith RC, Walsh K (1995) MyoD-induced expression of p21 inhibits cyclin-dependent kinase activity upon myocyte terminal differentiation. Mol Cell Biol 15:3823–3829

Halevy O, Novitch BG, Spicer DB, Skapek SX, Rhee J, Hannon GJ, Beach D, Lassar AB (1995) Correlation of terminal cell cycle arrest of skeletal muscle with induction of p21 by MyoD. Science 267:1018–1021

Hall M, Bates S, Peters G (1995) Evidence for different modes of action of cyclin-dependent kinase inhibitors: p15 and p16 bind to kinases, p21 and p27 bind to cyclins. Oncogene 11:1581–1588

Hannon GJ, Beach D (1994) p15INK4b is a potential effector of cell cycle arrest mediated by TGFβ. Nature 371:257–261

Harper JW, Adami GR, Wei N, Keyomarsi K, Elledge SJ (1993) The p21 CDK-interacting protein CIP1 is a potential inhibitor of G1 cyclin-dependent kinases. Cell 75:805–816

Hartwell LH, Kastan MB (1994) Cell cycle control and cancer. Science 266:1821–1828

Harvey M, McArthur MJ, Montgomery CA, Butel JS, Bradley A, Donehower LA (1994) Spontaneous and carcinogen-induced tumorigenesis in p53-deficient mice. Nature Genet 5:225–229

Hebert J, Cayuela JM, Berkeley J, Sigaux F (1994) Candidate tumor-suppressor genes MTS1 (p16^{INK4A}) and MTS2 (p15^{INK4B}) display frequent homozygous deletions in primary cells from T- but not from B-cell lineage acute lymphoblastic leukemias. Blood 84:4038–4044

Hengst L, Reed SI (1996) Translational control of p27Kip1 accumulation during the cell cycle. Science 271:1861–1864

Hennings H, Micahel D, Cheng C, Steinert P, Holbrook K, Yuspa SH (1980) Calcium regulation of growth and differentiation of mouse epidermal cells in culture. Cell 19:245–254

Hermeking H, Funk JO, Reichert M, Ellwart JW, Eick D (1995) Abrogation of p53-induced cell cycle arrest by c-Myc: evidence for an inhibitor of p21$^{WAF1/CIP1/SDI1}$. Oncogene 11:1409–1415

Herlyn M (1990) Human melanoma: development and progression. Cancer Metastas Rev 9:101–112

Hirai H, Roussel MF, Kato JY, Ashmun RA, Sherr CJ (1995) Novel INK4 proteins, p19 and p18, are specific inhibitors of the cyclin D-dependent kinases CDK4 and CDK6. Mol Cell Biol 15:2672–2681

Hollenstein M, Sidransky D, Vogelstein B, Harris CC (1991) p53 mutations in human cancers. Science, 253:49–53

Hollingsworth RJ, Hensey CE, Lee WH (1993) Retinoblastoma protein and the cell cycle. Curr Opin Genet Dev 3:55–62

Hooper ML (1994) The role of the p53 and Rb-1 genes in cancer, development and apoptosis. J Cell Sci Suppl 18:13–17

Horowitz JM, Udvadia AJ, (1995) Regulation of transcription by the retinoblastoma (Rb) protein. Mol Cell Diff 3(4):275–314

Huberman E, Callaham MF (1979) Induction of terminal differentiation in human promyeolcytic leukemia cells by tumor-promoting agents. Proc Natl Acad Sci USA 76:1293–1297

Hunter T, Pines J (1994) Cyclins and cancer II: cyclin D and CDK inhibitors come of age. Cell 79:573–582

Jacks T, Fazeli A, Schmitt EM, Bronson RY, Goodell MA, Weinberg RA (1992) Effects of an Rb mutation in the mouse. Nature 359:295–300

Jacks T, Remington L, Wiliams BO, Schmitt EM, Halachmi S, Bronson RT, Weinberg RA (1994) Tumor spectrum analysis in p53-mutant mice. Curr Biol 4:1–7

Jiang H, Fisher PB (1993) Use of a sensitive and efficient subtraction hybridization protocol for the identification of genes differentially regulated during the induction of differentiation in human melanoma cells. Mol Cell Diff 1(3):285–299

Jiang H, Su Z-Z, Boyd J, Fisher PB (1993) Gene expression changes associated with reversible growth suppression and the induction of terminal differentiation in human melanoma cells. Mol Cell Diff 1(1):41–66

Jiang H, Lin J, Fisher PB (1994a) A molecular definition of terminal cell differentiation in human melanoma cells. Mol Cell Diff 2(3):221–239

Jiang H, Lin J, Su Z-Z, Collart FR, Huberman E, Fisher PB (1994b) Induction of differentiation in human promyelocytic HL-60 leukemia cells activates p21, WAF1/CIP1, expression in the absence of p53. Oncogene 9:3389–3396

Jiang H, Lin JJ, Su Z-Z, Goldstein NI, Fisher PB (1995a) Subtraction hybridization identifies a novel melanoma differentiation associated gene, mda-7, modulated during human melanoma differentiation, growth and progression. Oncogene 11:2477–2486

Jiang H, Lin J, Su Z-Z, Herlyn M, Kerbel RS, Weissman BE, Welch DR, Fisher PB, (1995b) The melanoma differentiation associated gene mda-6, which encodes the cyclin-dependent kinase inhibitor p21, is differentially expressed during growth, differentiation and progression in human melanoma cells. Oncogene 10:1855–1864

Jiang H, Lin J, Young S-M, Goldstein NI, Waxman S, Davilla V, Chellappan SP, Fisher PB (1995c) Cell cycle gene expression and E2F transcription factor complexes in human melanoma cells induced to terminally differentiate. Oncogene 11:1179–1189

Jiang H, Lin J, Su Z-Z, Fisher PB (1996a) The melanoma differentiation associated gene-6 (mda-6), which encodes the cyclin-dependent kinase inhibitor p21, may function as a negative regulator of human melanoma growth and progression. Mol Cell Diff 4(1):67–89

Jiang H, Su Z-Z, Lin JJ, Goldstein NI, Young CSH, Fisher PB (1996b) The melanoma differentiation associated gene mda-7 suppresses cancer cell growth. Proc Natl Acad Sci USA 93:9160–9165

Jinno S, Suto K, Nagata A, Igarashi M, Kanaoka Y, Nojima H, Okayama H (1994) Cdc25A is a novel phosphatase functioning early in the cell cycle. EMBO J 13:1549–1556

Kacker RK, Giovanella BC, Pathak S (1990) Consistent karyotypic abnormalities in human malignant melanomas. Anticancer Res 10:859–872

Kamb A (1995) Cell-cycle regulators and cancer. Trends Genet 11:136–140

Kamb A, Gruis NA, Weaver-Feldhaus J, Liu Q, Harshman K, Tavtigian SV, Stockert E, Day RS III, Johnson BE, Skolnick MH (1994) A cell cycle regulator involved in genesis of many tumor types. Science 264:436–440

Kato J, Matsushime H, Hiebert SW, Ewen ME, Sherr CJ (1993) Direct binding of cyclin D to the retinoblastoma gene product (pRb) and pRb phosphorylation by the cyclin D-dependent kinase CDK4. Genes Dev 7:331–342

Kato JY, Matsuoka M, Polyak K, Massague J, Sherr CJ (1994a) Cyclic AMP-induced G1 phase arrest mediated by an inhibitor (p27Kip1) of cyclin-dependent kinase 4 activation. Cell 79:487–496

Kato JY, Matsuoka M, Strom DK, Sherr CJ (1994b) Regulation of cyclin D-dependent kinase 4 (cdk4) by cdk4-activating kinase. Mol Cell Biol 14:2713–2721

Kerbel RS (1990) Growth dominance and the metastatic cancer cell: cellular and molecular aspects. Adv Cancer Res 55:87–132

Kiyokawa H, Kineman RD, Manova-Todorova KO, Soares VC, Hoffman ES, Ono M, Khanam D, Hayday AC, Frohman LA, Koff A (1996) Enhanced growth of mice lacking the cyclin-dependent kinase inhibitor function of p27(Kip1). Cell 85:721–732

Kobayashi H, Man S, MacDougall JR, Graham CH, Lu C, Kerbel RS (1994) Variant sublines of early-stage human melanomas selected for tumorigenicity in nude mice express a multicytokine-resistant phenotype. Am J Pathol 144:776–786

Koff A, Ohtuski M, Polyak K, Roberts JM, Massague J (1993) Negative regulation of G1 in mammalian cells: Inhibition of cyclin E-dependent kinase by TGFβ. Science 260:536–539

Koh J, Enders GH, Dynlacht BD, Harlow E (1995) Tumour-derived p16 alleles encoding proteins defective in cell-cycle inhibition. Nature 375:506–510

Laiho M, DeCaprio J, Ludlow J, Livingston DM, Massague J (1990) Growth suppression by TGFβ linked to suppression of retinoblastoma protein phosphorylation. Cell 62:175–185

Lassar AB, Skapek SX, Novitch B (1994) Regulatory mechanisms that coordinate skeletal muscle differentiation and cell cycle withdrawal. Curr Opin Cell Biol 6:788–794

Lee EY-HP, Chang CY, Hu N, Wang Y-CJ, Lai C-C, Herrup K, Lee WH, Bradley A (1992) Mice deficient for Rb are nonviable and show defects in neurogenesis and hematopoiesis. Nature 359: 288–294

Lee MHJ, Reynisdottir I, Massague J (1995) Cloning of p57KIP2, a cyclin-dependent kinase inhibitor with unique domain structure and tissue distribution. Genes Dev 9:639–649

Lee MH, Nikolic M, Baptista CA, Lai E, Tsai LH, Massague J (1996) The brain-specific activator p35 allows Cdk5 to escape inhibition by p27KIP1 in neurons. Proc Natl Acad Sci USA 93:3259–3263

Li R, Waga S, Hannon GJ, Beach D, Stillman B (1994) Differential effects by the p21 CDK inhibitor on PCNA-dependent DNA replication and repair. Nature 371:534–537

Li S, MacLachlan TK, De Luca A, Claudio PP, Condorelli G, Giordano A (1995) The cdc-2-related PISSLRE, is essential for cell growth and acts in G2 phase of the cell cycle. Cancer Res 55:3992–3996

Li Y, Jenkins CW, Nichols MA, Xiong Y (1994) Cell cycle expression and p53 regulation of the cyclin-dependent kinase inhibitor p21. Oncogene 9:2261–2268

Liotta LA, Steeg PG, Stetler-Stevenson WG (1991) Cancer metastasis and angiogenesis: an imbalance of positive and negative regulation. Cell 64:327–336

Liu M, Lee M-H, Cohen M, Bommakanti M, Freedman LP (1996) Transcriptional activation of the Cdk inhibitor p21 by vitamin D_3 leads to the induced differentiation of the myelomonocytic cell line U937. Genes Dev 10:142–153

Loganzo F Jr, Nabeya Y, Maslak P, Albino AP (1994) Stabilization of p53 protein is a critical response to UV radiation in human melanocytes: implications for melanoma development. Mol Cell Diff 2(1): 23–43

Lotem J, Sachs L (1979) Regulation of normal differentiation in mouse and human myeloid leukemic cells by phorbol esters and the mechanism of tumor promotion. Proc Natl Acad Sci USA 76:5158–5162

Lu C, Kerbel RS (1994) Cytokines, growth factors and the loss of negative growth controls in the progression of human cutaneous malignant melanoma. Curr Opin Oncol 6:212–220

Lukas J, Parry D, Aagaard L, Mann DJ, Bartkova J, Strauss M, Peters G, Bartek J (1995) Retinoblastoma-protein-dependent cell-cycle inhibition by the tumour suppressor p16. Nature 375:503–506

Luo Y, Hurwitz J, Massague, J (1995) Cell-cycle inhibition by independent CDK and PCNA binding domains in $p21^{Cip1}$. Nature 375:159–161

MacDougall JR, Kobayashi H, Kerbel RS (1993) Responsiveness of normal/dysplastic melanocytes and melanoma cells from different lesional stages of disease progression to the growth inhibitory effects of TGF-ß. Mol Cell Diff 1(1):21–40

MacLachlan TK, Sang N, Giordano A (1995) Cyclins, cyclin-dependent kinases and Cdk inhibitors: implications in cell cycle control and cancer. Crit Rev Eukaryot Gene Express 5(2):127–156

Macleod FF, Sherry N, Hannon G, Beach D, Tokino T, Kinzler K, Vogelstein B, Jacks T (1995) p53-dependent and independent expression of p21 during cell growth, differentiation and DNA damage. Genes Dev 9:935–944

MacJohnson DD, Vojta PJ, Barrett JC, Noda A, Pereira-Smith OM, Smith JR (1994) Evidence for a p53-independent pathway for upregulation of SDI1/CIP1/WAF1/p21 RNA in human cells. Mol Carcinog 11:59–64

Mal A, Poon RY, Howe PH, Toyoshima H, Hunter T, Harter ML (1996) Inactivation of p27Kip1 by the viral E1A oncoprotein in TGFbeta-treated cells. Nature 380:262–265

Mangelsdorf DJ, Pike JW, Haussler MR (1984) 1,25-dihydroxyvitamin D_3-induced differentiation in a human promyelocytic leukemia cell line (HL-60): receptor-mediated maturation to macrophage-like cells. J Cell Biol 98:391–398

Massague J, Polyak K (1995) Mammalian antiproliferative signals and their targets. Curr Opin Genet Dev 5:91–96

Marshall CJ (1991) Tumor suppressor genes. Cell 64:313–326

Matsuoka M, Kato JY, Fisher RP, Morgan DO, Sherr CJ (1994) Activation of cyclin-dependent kinase 4 (cdk4) by mouse MO15-associated kinase. Mol Cell Biol 14:7265–7275

Matsuoka S, Edwards MC, Bai C, Parker S, Zhang P, Baldini A, Harper JW, Elledge SJ (1995) p57KIP2, a structurally distinct member of the p21CIP1 Cdk inhibitor family, is a candidate tumor suppressor gene. Genes Dev 9(6):650–662

Matsushime H, Roussel M, Ashmun R, Sherr CJ (1991) Colony stimulating factor 1 regulates novel cyclins during the G1 phase of the cell cycle. Cell 65:701–713

Matsushime H, Quelle DE, Shurtleff SA, Shibuya M, Sherr CJ, Kato JY (1994) D-type cyclin-dependent kinase activity in mammalian cells. Mol Cell Biol 14:2066–2076

McCarthy DM, San Miguel J, Freake HC, Green PM, Zola H, Catowsky D, Goldman J (1983) 1,25-dihydroxyvitamin D_3 inhibits the proliferation of human promyelocytic leukemia cell line (HL-60) cells and induces monocyte-macrophage differentiation in HL-60 and normal bone marrow cells. Leuk Res 7:51–55

Michieli P, Chedid M, Lin D, Pierce JH, Mercer WE, Givol D (1994) Induction of WAF1/CIP1 by a p53-independent pathway. Cancer Res 54:3391–3395

Miele ME, Robertson G, Lee J-H, Coleman A, McGary CT, Fisher PB, Lugo TG, Welch DR (1996) Metastasis suppressed, but tumorigenicity and local invasiveness unaffected, in the human melanoma cell line MelJuSo after introduction of human chromosome 1 or 6. Mol Carcinog 15:284–299

Missero C, Calautti E, Eckner R, Chin J, Tsai LH, Livingston DM, Dotto PG (1995) Involvement of the cell-cycle inhibitor Cip1/Waf1 and the E1A-associated p300 protein in terminal differentiation. Proc Natl Acad Sci USA 92:5451–5455

Missero M, Dotto GP (1996) p21$^{WAF1/CIP1}$ and terminal differentiation control of normal epithelia. Mol Cell Diff 4(1):1–16

Miyaura C, Abe E, Kuribayashi T, Tanaka H, Konno K, Nishii Y, Suda T (1981) 1α,25-dihydroxyvitamin D_3 induces differentiation of human myeloid leukemia cells. Biochem Bipophys Res Commun 102:937–943

Montano X, Shamsher M, Whitehead P, Dawson K, Newton J (1994) Analysis of p53 in human cutaneous melanoma cell lines. Oncogene 9:1455–1459

Morgan DO (1992) Cell cycle control in normal and neoplastic cells. Curr Opin Genet Dev 2:33–37

Morgan DO (1995) Principles of CDK regulation. Nature 374:131–134

Morgan DO (1996) Under arrest at atomic resolution. Nature 382:295–296

Morgan DO, De Bondt HL (1994) Protein kinase regulation: insights from crystal structure analysis. Curr Opin Cell Biol 6(2):239–246

Murao S-I, Gemmell MA, Callaham MF, Anderson NL, Huberman E (1983) Control of macrophage cell differentiation in human promyelocytic HL-60 leukemia cells by 1,25-dihydroxyvitamin D_3 and phorbol-12-myristate-13-acetate. Cancer Res 43:4989–4996

Nakanishi M, Robetorge RS, Adami GR, Pereira-Smith OM, Smith JR (1995) Identification of the active region of the DNA synthesis inhibitory gene p21$^{Sdi1/CIP1/WAF1}$. EMBO J 14:555–563

Nakayama K, Ishida N, Shirane M, Inomata A, Inoue T, Shishido N, Horii I, Loh DY, Nakayama K (1996) Mice lacking p27(Kip1) display increased body size, multiple organ hyperplasia, retinal dysplasia, and pituitary tumors. Cell 85:707–720

Nevins JR (1992) E2F: a link between the Rb tumor suppressor protein and viral oncoproteins. Science 258:424–429

Nobori T, Miura K, Wu DJ, Lois A, Takabayashi K, Carson DA (1994) Deletions of the cyclin-dependent kinase 4 inhibitor gene in multiple human cancers. Nature 368:753–756

Noda A, Ning Y, Venable SF, Pereira-Smith OM, Smith JR (1994) Cloning of senescent cell-derived inhibitors of DNA synthesis using an expression screen. Exp Cell Res 211:90–98

Nourse J, Firpo E, Flanagan WM, Coats S, Polyak K, Lee MH, Massague J, Crabtree J, Roberts JM (1994) Interleukin 2-mediated elimination of p27KIP1 cyclin-dependent kinase inhibitor prevented by rapamycin. Nature 372:570–573

Okamoto A, Demetrick DJ, Spillare EA, Hagiwara K, Hussain SP, Bennett WP, Forrester K, Gerwin B, Serrano M, Beach DH, Harris CC (1994) Mutations and altered expression p16INK4 in human cancer. Proc Natl Acad Sci USA 91:11045–11049

Okuda T, Hirai H, Valentine VA, Shurtleff SA, Kidd VJ, Lahti JM, Sherr CJ, Downing JR (1995a) Molecular cloning, expression pattern, and chromosomal localization of human CDKN2D/INK4d, an inhibitor of cyclin D-dependent kinases. Genomics 29:623–630

Okuda T, Shurtleff SA, Valentine MB, Raimondi SC, Head DR, Behm F, Curcio-Brint AM, Liu Q, Pui C-H, Sherr CJ, Beach D, Look AT, Downing JR (1995b) Frequent deletion of p16INK4a/MTS1 and p15INK4b/MTS2 in pediatric acute lymphoblastic leukemia. Blood 85:2321–2330

Olson EN, Klein WH (1994) bHLH factors in muscle development: dead lines and commitments, what to leave in and what to leave out. Genes Dev 8:1–8

Pagano M, Tam SW, Theodoras AM, Beer-Romero P, Del Sal G, Chau V, Yew PR, Draetta GF, Rolfe M (1995) Role of the ubiquitin-proteasome pathway in regulating abundance of the cyclin-dependent kinase inhibitor p27. Science 269:682–685

Paggi MG, Baldi A, Bonetto F, Giordano A (1996) The retinoblastoma protein family in cell cycle and cancer. J Cell Biochem 62:418–430

Parker SB, Eichele G, Zhang P, Rawls A, Sands AT, Bradley A, Olson EN, Harper JW, Elledge SJ (1995) p53-independent expression of p21^{CIP1} in muscle and other terminally differentiating cells. Science 267:1024–1027

Parry D, Bates S, Mann DJ, Peters G (1995) Lack of cyclin D-Cdk complexes in Rb-negative cells correlates with high levels of p16INK4/MTS1 tumour suppressor gene product. EMBO J 14(3):503–511

Pathak S, Drwinga HL, Hsu TC (1983) Involvement of chromosome 6 in rearrangements in human malignant melanoma. Cancer Genet Cytotgenet 36:573–579

Picksley SM, Lane DP (1994) p53 and Rb: their cellular roles. Curr Opin Cell Biol 6:853–858

Pientenpol JA, Bohlander SK, Sato Y, Papadopoulos N, Liu B, Friedman C, Trask BJ, Roberts JM, Kinzler KW, Rowley JD, Vogelstein B (1995) Assignment of the human p27Kip1 gene to 12p13 and its analysis in leukemias. Cancer Res 55:1206–1210

Pines J (1995a) Cyclins and cyclin-dependent kinases: theme and variations. Adv Cancer Res 66:181–212

Pines J (1995b) Cyclins, CDKs and cancer. Semin Cancer Biol 6:63–72

Pines J (1995c) Cyclins and cyclin-dependent kinases: a biochemical view. Biochem J 308:697–711

Polyak K, Kato JY, Solomon MJ, Sherr CJ, Massague J, Roberts JM, Koff A (1994a) p27Kip1, a cyclin-Cdk inhibitor, links transforming growth factor-beta and contact inhibition to cell cycle arrest. Genes Dev 8:9–22

Polyak K, Lee M-H, Erdjument-Bromage H, Koff A, Roberts JM, Tempst P, Massague J (1994b) Cloning of p27KIP1, a cyclin-dependent-kinase inhibitior and a potential mediator of extracellular antimitogenic signals. Cell 78:59–76

Ponce-Castaneda MV, Lee MH, Latres E, Polyack K, Lacombe L, Montgomery K, Mathew S, Krauter K, Massague J, Cordon-Cardo C (1995) p27^{Kip1}: chromosomal mapping to 12p12 and absence of mutations in human tumors. Cancer Res 55:1211–1214

Poon RY, Hunter T (1995) Cell regulation. Innocent bystanders or chosen collaborators? Curr Biol 5:1243–1247

Poon RY, Yamashita K, Adamczewski JP, Hunt T, Shuttleworth J (1993) The cdc2-related protein p40MO15 is the catalytic subunit of a protein kinase that can activate cdc2. EMBO J 12(8):3123–3132

Poon RY, Toyoshima H, Hunter T (1995) Redistribution of the CDK inhibitor p27 between different cyclin-CDK complexes in the mouse fibroblast cell cycle and in cells arrested with lovastatin or ultraviolet irradiation. Mol Biol Cell 6:1197–1213

Quelle DE, Zindy F, Ashmun RA, Sherr CJ (1995a) Alternative reading frames of the INK4a tumor suppressor gene encode two unrelated proteins capable of inducing cell cycle arrest. Cell 83:993–1000

Quelle DE, Ashmun RA, Hannon GJ, Rehberger PA, Trono D, Richter KH, Walker C, Beach D, Sherr CJ, Serrano M (1995b) Cloning and characterization of murine p16INK4a and p15INK4b genes. Oncogene 11:635–645

Rao SS, Chu C, Kohtz DS (1994) Ectopic expression of cyclin D1 prevents activation of gene transcription by myogenic basic helix-loop-helix regulators. Mol Cell Biol 14:5259–5267

Reddy GP (1994) Cell cycle: regulatory events in G1→S transition of mammalian cells. J Cell Biochem 54:379–386

Reed SI, Bailly E, Dulic V, Hengst L, Resnitzky D, Slingerland J (1994) G1 control in mammalian cells. J Cell Sci Suppl 18:69–73

Reynisdottir I, Polyak K, Iavarone A, Massague J (1995) Kip/Cip and Ink4 Cdk inhibitors cooperate to induce cell cycle arrest in response to TGF-beta. Genes Dev 9:1831–1845

Robetorye RS, Nakanishi M, Venable SF, Pereira-Smith OM, Smith JR (1996) Regulation of p21$^{Sdi1/Cip1/Waf1/mda-6}$ and expression of other cyclin-dependent kinase inhibitors in senescent cell. Mol Cell Diff 4(1):113–126

Rovera G, Santoli D, Damsky C (1979) Human promyeolcytic leukemia cells in culture differentiate into macrophage-like cells when treated with a phorbol diester. Proc Natl Acad Sci USA 76:2779–2783

Russo AA, Jeffrey PD, Patten AK, Massague J, Pavletich NP (1996) Crystal structure of the p27Kip1 cyclin-dependent-kinase inhibitor bound to the cyclin A-cdk2 complex. Nature 382:325–331

Sang N, Baldi A, Giordano A (1995) The roles of tumor suppressors pRb and p53 in cell proliferation and cancer. Mol Cell Diff 3(1):1–29

Schneider JW, Gu W, Zhu L, Mahdavi V, Nadal-Ginard B (1995) Reversal of terminal differentiation meidated by p107 in Rb−/− muscle cells. Science 264:1467–1471

Schwaller J, Koeffler HP, Niklaus G, Loetscher P, Magel S, Fey MF, Tobler A (1995) Posttranscriptional stabilization underlies p53-independent induction of p21/WAF1/CIP1/SDI1 in differentiating human leukemic cells. J Clin Invest 95:973–979

Serrano M, Hannon GJ, Beach D (1993) A new regulatory motif in cell cycle control causing specific inhibition of cyclinD-cdk4. Nature 366:704–707

Serrano M, Gomez-Lahoz E, DePinho RA, Beach D, Bar-Sagi D (1995) Inhibition of ras-induced proliferation and cellular transformation by p16INK4. Science 267:249–252

Serrano M, Lee H, Chin L, Cordon-Cardo C, Beach D, DePinho RA (1996) Role of INK4a locus in tumor suppression and cell mortality. Cell 85:27–37

Sheikh MS, Li X-S, Chen J-C, Shao Z-M, Ordonez JV, Fontana JA (1994) Mechanisms of regulation of WAF1/Cip1 gene expression in human breast carcinoma: role of p53-dependent and independent signal transduction pathways. Oncogene 9:3407–3415

Sherr CJ (1993) Mammalian G1 cyclins. Cell 73:1059–1065

Sherr CJ (1994a) G1 phase progression: cycling on cue. Cell 79:551–555
Sherr CJ (1994b) Growth-factor-regulated G1 cyclins. Stem Cells 12[Suppl] 1:47–55
Sherr CJ (1995) D-type cyclins. Trends Biochem. Sci 20:187–190
Sherr CJ, Roberts JM (1995) Inhibitors of mammalian G_1 cyclin-dependent kinases. Genes Dev 9:1149–1163
Sherr CJ, Kato J, Quelle DE, Matsuoka M, Roussel MF (1994) D-type cyclins and their cyclin-dependent kinases: G1 phase integrators of the mitogenic response. Cold Spring Harb Symp Quant Biol 59:11–19
Shiekhattar R, Mermelstein F, Fisher, RP, Drapkin R, Dynlacht B, Wessling HC, Morgan DO, Reinberg D (1995) Cdk-activating kinase complex is a component of human transcription factor TFIIH. Nature 374:283–287
Shivji MKK, Grey SJ, Strausfeld UP, Wood RD, Blow JJ (1994) Cip1 inhibits DNA replication but not PCNA-dependent nucleotide excision-repair. Curr Biol 4:1062–1068
Skapek SX, Rhee J, Spicer DB, Lassar AB (1995) Inhibition of myogenic differentiation in proliferating myoblasts by cyclin D1-dependent kinase. Science 267:1022–1027
Smith JR, Pereira-Smith OM (1996) Replicative senescence: implications for in vivo aging and tumor suppression. Science 273:63–67
Steinman RA, Hoffman B, Iro A, Guillouf C, Liebermann DA, El-Houseini ME (1994) Induction of p21 (WAF1/CIP1) during differentiation. Oncogene 9:3389–3396
Stone S, Jiang P, Dayananth P, Tavtigian SV, Katcher H, Parry D, Peters G, Kamb A (1995) Complex structure and regulation of the P16 (MTS1) locus. Cancer Res 55:2988–2994
Su Z-Z, Austin VA, Zimmer SG, Fisher PB (1993) Defining the critical gene expression changes associated with expression and suppression of the tumorigenic and metastatic phenotype in Ha-ras-transformed cloned rat embryo fibroblast cells. Oncogene 8:1211–1219
Su Z-Z, Yemul S, Estabrook A, Zimmer SG, Friedman RM, Fisher PB (1995) Transcriptional switching model for the regulation of tumorigenesis and metastasis by the Ha-ras oncogene: transcriptional changes in the Ha-ras tumor suppressor gene lysyl oxidase. Intl J Oncology 7:1279–1284
Tam SW, Shay JW, Pagano M (1994) Differential expression and cell cycle regulation of the cyclin dependent kinase 4 inhibitior p16INK4. Cancer Res 54:5816–5820
Toyoshima H, Hunter T (1994) p27, a novel inhibitior of G1-cyclin-cdk protein kinase activity is related to p21. Cell 78:67–74
Trent JM (1991) Cytogenetics of human malignant melanoma. Cancer Metastas Rev 10:103–113
Trent JM, Stanbridge EJ, McBride HL, Meese EU, Casey G, Araujo DE, Witkowski CM, Nagle RB (1990) Tumorigenicity in human melanoma cell lines controlled by introduction of human chromosome 6. Science 247:568–571
Ueki K, Ono Y, Henson JW, Efird JT, von Deimling A, Louis DN (1996) CDKN2/p16 or RB alterations occur in the majority of glioblastomas and are inversely correlated. Cancer Res 56:150–153
Volkenandt M, Schlegel U, Nanus DM, Albino AP (1991) Mutational analysis of the human p53 gene in malignant melanoma. Pigment Cell Res 4:35–40
Waga S, Hannon GJ, Beach D, Stillman B (1994) The p21 inhibitor of cyclin-dependent kinases controls DNA replication by interaction with PCNA. Nature 369:574–578
Walsch K, Guo K, Wang J, Andres V (1996) p21 Regulation and function during myogenesis. Mol Cell Diff 4(1):17–31
Wang J, Walsh K (1996) Resistance to apoptosis conferred by Cdk inhibitors during myocyte differentiation. Science 273:359–361
Weinberg RA (1995) The retinoblastoma protein and cell cycle control. Cell 81:323–330
Weintraub H (1993) The MyoD family and myogenesis: redundancy, networks and thresholds. Cell 75:1241–1244
Welch DR, Rieber M (1996) Is $p21^{WAF1}$ a suppressor of malignant melanoma metastasis? Mol Cell Diff 4(1):91–111
Welch DR, Bisi JE, Miller BE, Conaway D, Seftor EA, Gilmore LB, Seftor REB, Nakajima M, Hendrix MJC (1991) Characterization of a highly invasive and spontaneously metastatic human malignant melanoma cell line. Int J Cancer 47:227–237
Welch DR, Chen P, Miele ME, McGary CT, Bower JM, Weissman BE, Stanbridge EJ (1994) Microcell-mediated transfer of chromosome 6 into metastatic human C8161 melanoma cells suppresses metastasis but does not inhibit tumorigenicity. Oncogene 9:255–262
Whyte P (1995) The retinoblastoma protein and its relatives. Semin Cancer Biol 6:83–90
Williams BO, Schmitt, EM, Remington L, Bronson RT, Albert DM, Weinberg RA, Jacks T (1994a) Extensive contribution of Rb deficient cells to adult chimeric mice with limited histopathological consequences. EMBO J 13:4251–4259

Williams BO, Remington L, Albert DM, Mukai S, Bronson RT, Jacks T (1994) Cooperative tumorigenic effects of germ-line mutations in Rb ahd p53. Nature Genet 7:480–484

Wu H, Wade M, Krall L, Grisham J, Xiong Y, Van Dyke T (1996) Targeted in vivo expression of the cyclin-dependent kinase inhibitor p21 halts hepatocyte cell-cycle progression, postnatal liver development, and regeneration. Genes Dev 10:245–260

Xiong Y, Zhang H, Beach D (1992) D-type cyclins associate with multiple protein kinases and the DNA replication and repair factor PCNA. Cell 71:505–514

Xiong Y, Hannon GJ, Zhang H, Casso D, Kobayashi R, Beach D (1993a) p21 is a universal inhibitor of cyclin kinases. Nature 366:701–704

Xiong Y, Zhang H, Beach D (1993b) Subunit rearrangement of the cyclin-dependent kinases is associated with cellular transformation. Genes Dev 7:1572–1583

Yang Z-Y, Perkins ND, Ohno O, Nabel EG, Nabel GJ (1995) The p21 cyclin-dependent kinase inhibitor suppresses tumorigenicity in vivo. Nature Med 1:1052–1056

Yuan W, Condorelli G, Caruso M, Felsani A, Giordano A (1996) Human p300 protein is a coactivator for the transcription factor MyoD. J Biol Chem 271:9009–9013

Zakut R, Givol D (1995) The tumor suppressor function of $p21^{Waf}$ is contained in its N-terminal half ('half-WAF'). Oncogene 11:393–395

Zhang H, Xiong Y, Beach D (1993) Proliferating cell nuclear antigen and p21 are components of multiple cell cycle kinase complexes. Mol Biol Cell 4:897–906

Zhang H, Hannon GJ, Beach D (1994) p21-containing cyclin kinases exist in both active and inactive states. Genes Dev 8:1750–1758

Zhang W, Grasso L, McClain CD, Gambel AM, Cham Y, Travali S, Deisseroth AB, Mercer WE (1995) p53-independent induction of WAF1/CIP1 in human leukemia cells is correlated with growth arrest accompanying monocyte/macrophage differentiation. Cancer Res 55:668–674

Roles of Cyclin-Dependent Kinase Inhibitors: Lessons from Knockout Mice

H. KIYOKAWA and A. KOFF

1	Introduction.	105
2	Creation of Knockout Mice	105
3	Analyses of the In Vivo Expression of the CDK Inhibitors	107
4	Checkpoint Protein p21^{Cip1}	108
5	Functions of p27^{Kip1}	109
5.1	Growth-Regulatory Function	109
5.2	Thymic Development	111
5.3	Guardian of Quiescence	112
5.4	Female Fertility	113
5.5	Retinal Development	114
6	p16^{Ink4A}, a Bona Fide Tumor Supperssor	115
7	Putative Physiological Roles of the CDK Inhibitors	117
References		118

1 Introduction

In the field of molecular genetics, analysis of mutant organisms has long been a powerful tool to understand the physiological function of a gene. Owing to the development of sophisticated vector systems, in combination with the establishment of mouse embryonic stem (ES) cells, targeted mutagenesis of a mouse gene is now widely used to "knock out" any gene of interest to obtain knowledge of its function. This review focuses on the approaches using knockout mice to understand the physiological function of the cyclin-dependent kinase (CDK) inhibitors.

2 Creation of Knockout Mice

Two of the most important conditions to successfully generate a knockout mouse strain are the establishment of an ES cell culture from mouse embryos and the

---

Molecular Biology Program, Memorial Sloan-Kettering Cancer Center, 1275 York Avenue, New York, NY 10021, USA

appropriate choice of recombination vector. ES cells are derived from mouse embryos 3.5-days postcoitum and arise from the inner cell mass of the blastocyst (ABBONDANZO et al. 1993). The ability to culture ES cells for many passages without loss of totipotentiality allows these cells to contribute to virtually all the tissues of an animal when injected back into a host blastocyst. In addition, the appropriate choice of vector for homologous recombination can reduce the time required to screen for ES clones carrying the mutation of interest. The technique of positive/negative selection, pioneered by Cappecchi and colleagues (MANSOUR et al. 1988), has gained wide acceptance, while other strategies such as "promoter-less" vectors and "polyA-less" vectors have their own advantages, as discussed elsewhere (RAMIREZ-SOLIS et al. 1993).

Positive/negative selection relies on isolating ES cells resistant to both G418 and ganciclovir. The targeting vector contains a neomycin resistance (*neo*) gene within homologous regions of the target sequence and a herpes simplex virus thymidine kinase (*TK*) gene outside these regions. The construct we used to target the p27^{Kip1} gene is shown in Fig. 1. Resistance to G418 indicates the presence of the *neo* gene. Resistance to ganciclovir indicates the absence of the *TK* gene. Other genes with similar positive or negative selection features can be used. If site-specific recombination occurs, then integration of *neo* and elimination of *TK* will be observed. In contrast, if random integration occurs, then both *neo* and *TK* could be integrated. Legitimate homologous recombination in the selected ES clones is later confirmed by Southern blotting or polymerase chain reaction (PCR). Although the use of positive/negative selection does not eliminate the selection of false positives, it can facilitate the screening procedure five- to tenfold. Following transfection and selection, the appropriately integrated target creates an ES clone with both mutant and wild-type alleles of the target gene, if the locus is on a somatic chromosome.

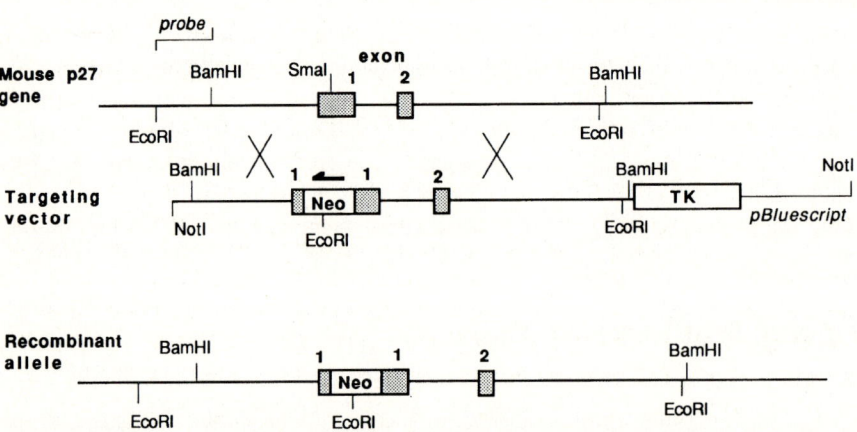

Fig. 1. Structures of the mouse p27^{Kip1} gene and a targeting vector for homologous recombination. The coding exons are shown as *closed boxes*, and the region used as a probe for Southern analysis is indicated. *Neo*, neomycin resistance gene; *TK*, herpes simplex virus thymidine kinase gene. The inserted Neo gene has the opposite transcriptional direction to the p27^{Kip1} gene

The contribution of injected ES cells to the germ cells of chimeric animals is the key to the establishment of a mutant mouse strain. The contribution of ES cells to the chimera depends on various factors (for a review, see ABBONDANZO et al. 1993), including the strain, state of sex- chromosomes, and culture condition of ES cells as well as the strain and preparation of host blastocysts. Most widely used ES lines have an XY genotype, because male ES cells tend to dominate the development of chimeric animals when injected into a female blastocyst. To facilitate analysis of chimeras, donor ES cells and recipient host blastocysts are derived from strains of mice that differ in coat color. Typically, ES cells derived from homozygous agouti 129/Sv mice are introduced into host C57BL/6 embryos that are homozygous for black coat color, although combinations of other strains of mice can be used. Chimeric animals have mixed coat color, because they contain cells originating from both the ES donor and the embryo recipients. Extensive chimeras, those mice with a large contribution of agouti ES cells, are likely to have substantial numbers of germ cells contributed by the injected ES cells. The ability to transmit the information in the agouti ES cell is assessed by crossing chimeric males with C57BL/6 females. Because agouti is dominant to black, agouti offspring indicate that ES cell-derived genetic information is transmitted through the germ line. Genotypes of the agouti offspring are subsequently determined by Southern blot or PCR. Homozygous mice are obtained from intercross breeding of heterozygous mice. It is important to note that these mice have a hybrid genetic background, in this case 129/Sv × C57BL/6, which might affect penetrance of phenotypes in the mutant animal.

3 Analyses of the In Vivo Expression of the CDK Inhibitors

There are two gene families of CDK inhibitors. The Cip/Kip gene family currently consists of three members, i.e., $p21^{Cip1/Waf1/Sdi1/Cap20}$, $p27^{Kip1}$, and $p57^{Kip2}$, and the Ink4 family has four members, i.e., $p16^{Ink4A/MTS1}$, $p15^{Ink4B/MTS2}$, $p18^{Ink4C}$, and $p19^{Ink4D}$ (for a review, see SHERR and ROBERTS 1995). The Cip/Kip proteins inhibit all G_1-phase CDK, whereas the Ink4 proteins are specific inhibitors of cyclin D-dependent kinases CDK4 and CDK6. Although other chapters cover the biochemistry and cell biology of these inhibitors, we will mention the essential points related to the in vivo roles of these proteins. To define the in vivo function of a gene, one of the critical studies is to determine the expression of the gene product in the wild-type animal. As organs are composed of many different types of cells, it is important to examine the expression histologically. Published studies of the CDK inhibitors using in situ hybridization are limited to p21 (PARKER et al. 1995) and p57 (MATSUOKA et al. 1995). During mouse embryogenesis, both p21 and p57 mRNA are highly expressed in dermomyotome at gestational days 8.5–10. High expression of p21 mRNA was detected in muscle, cartilage, epidermis, and nasal epithelial cells of mouse embryos. p57 mRNA is expressed highly in muscle, cartilage, brain, and lens epithelium of the eye during embryogenesis. In adult mice, p21 is highly expressed in liver, muscle, and the epithelium of intestine and stomach, while p57 expression is observed in kidney, muscularis mucosae in stomach, and muscle. In general, the expression of p21 and p57 is correlated with exit from the cell cycle during terminal differentiation, although these two molecules have distinct patterns of expression in most tissues.

In many cell lines, the abundance of p27 mRNA does not reflect the abundance of p27 protein (Soos et al. 1996). Consequently, it is important to examine the tissue distribution of p27 protein rather than its mRNA. We and others detected p27 protein in extracts derived from all organs of adult mice (Kiyokawa et al. 1996; Fero et al. 1996; Nakayama et al. 1996). However, the amount of p27, on a per microgram basis of the extract, varied widely. The amount of p27 was the highest in the thymus and spleen, tissues in which most of the cells are undergoing post partum cell fate determination. Immunohistochemistry has shown that, during thymic development, p27 protein level decreases as thymocytes resume proliferation following successful recombination of the T cell receptor (TCR)-β chain (Hoffman et al. 1996). In the ovary, the amount of p27 protein markedly increases as the granulosa cells in the follicle differentiate into the luteal cells with cessation of proliferation (Kiyokawa et al. 1996). These observations are consistent with the hypothesis that p27 regulates entry into and withdrawal from the cell cycle during post partum determination of cell fate.

4 Checkpoint Protein p21^{Cip1}

In wild-type mice, a strong correlation has been shown between the expression of p21 mRNA and terminal differentiation during embryonic development (Parker et al. 1995). Two additional observations further suggest a role for p21 in muscle differentiation; first, introduction of the muscle-specific transcription factor myoD into nonmyogenic cells induces muscle differentiation and a concomitant increase in the transcription of p21 (Halevy et al. 1995); second, myoblast differentiation is prevented by ectopic cyclin D expression, which may activate CDK4 beyond the inhibitory threshold set by p21 (Skapek et al. 1995). Involvement of p21 in differentiation is also supported by observations that introduction of p21 induces differentiation in neurogenic and hematopoietic cell lines (Kranenburg et al. 1995; Liu et al. 1996). However, mice lacking p21 develop normally (Brugarolas et al. 1995; Deng et al. 1995). This might be attributed to functional redundancy, although the redundant molecules have not yet been determined.

The lack of a phenotype in p21-knockout mice does not suggest a developmental role for the protein. It has been speculated that p21-mediated growth arrest might be part of a p53-dependent checkpoint (El-Deiry et al. 1993; Dulic et al. 1994). Consistent with a checkpoint function, embryonic fibroblasts prepared from p21-knockout mice fail to completely arrest in G_1 phase in response to DNA damage and nucleotide pool perturbation (Brugarolas et al. 1995; Deng et al. 1995). Moreover, p21-deficient fibroblasts, as well as p53-deficient fibroblasts, reach higher saturation density than wild-type cells, especially at late passages. These data suggest that p21 mediates p53-dependent growth arrest upon DNA damage and nucleotide pool perturbation and plays a role in regulation of growth kinetics of fibroblasts in culture. However, further comparison of the phenotypes of p53$^{-/-}$ and p21$^{-/-}$ mice indicates that p21 does not mediate all the checkpoint functions of p53. In fact, other p53-dependent pathways, such as apoptosis and the mitotic spindle checkpoint (CROSS et al. 1995), are not affected in p21-deficient mice.

Theoretically, any CDK inhibitor may have a tumor suppressor function by keeping quiescent cells from inappropriate reentry into the mitotic cycle. p21-knockout mice do not display increased spontaneous tumorigenesis, whereas p53-knockout mice have a rapid rate of early tumor formation (DONEHOWER et al. 1992). Consistent with the lack of tumorigenesis in p21-deficient mice, there is little evidence for the involvement of p21 in human tumors (SHIOHARA et al. 1994; for a review, see HIRAMA and KOEFFLER 1995) except a point mutation that results in heterozygosity in a single Burkitt's lymphoma (BHATIA et al. 1995). Nonetheless, keratinocytes obtained from p21-knockout mice are highly sensitive to Ras-induced transformation, and these tumors are much more aggressive than those from wild-type keratinocytes with Ras (MISSERO et al. 1996), suggesting that p21 deficiency could contribute to promotion of malignant tumorigenesis.

5 Functions of p27^{Kip1}

Characterization of various primary cells and tumor-derived cell lines suggested that p27 may play a critical role in regulating entry into and exit from the mitotic cycle in response to extracellular signals. There is a correlation between growth arrest and an increase in the CDK2 inhibitory activity of p27 either by an increase in p27 protein or release of p27 from sequestration in the cyclin D–CDK complex (for a review, see KOFF and POLYAK 1995). As part of an inhibitory threshold, p27 may antagonize the S phase-promoting effect of CDK and affect the decision to proliferate or withdraw from the mitotic cycle. Under in vivo conditions, mitogenic and antimitogenic signals rarely affect the cells in an "all-or-none" fashion. In most cases, there is crosstalk between signal-transduction pathways, which converge at the cell-cycle machinery. At this point in G_1 phase of the cell cycle, the restriction point, a cell makes a decision whether to enter S phase or withdraw from the cycle (see Fig. 6). This stochastic event depends on the G_1-phase CDK activity in each cell, which reflects the expression of cyclins, CDK, and CDK inhibitors as well as modifications of all of these proteins.

5.1 Growth-Regulatory Function

The creation of p27-deficient mice has revealed multiple functions of p27 in development (KIYOKAWA et al. 1996; FERO et al. 1996; NAKAYAMA et al. 1996). The

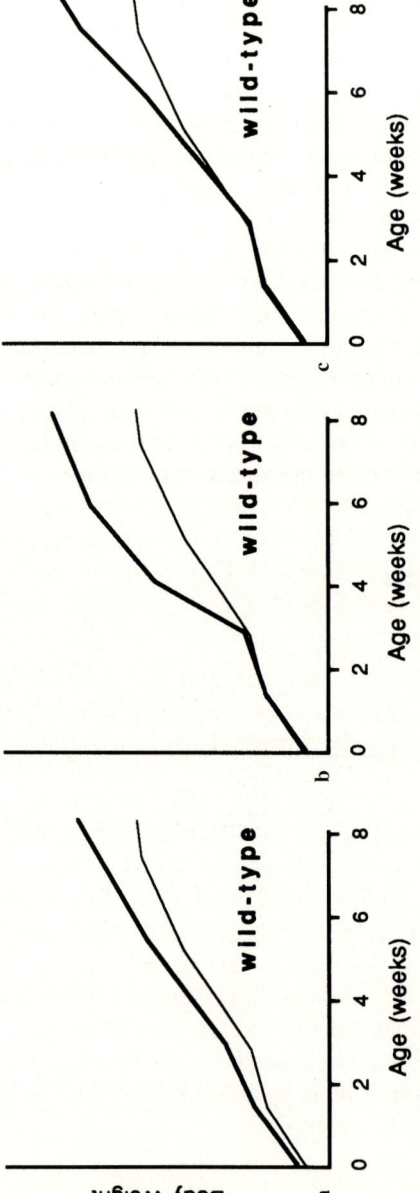

Fig. 2. Representative growth curves of **a** $p27^{Kip1}$-knockout mice, **b** growth hormone-transgenic mice, and **c** insulin-like growth factor-1 transgenic mice in comparison with wild-type mice

most obvious phenotype is that p27-deficient mice grow 20%–40% larger than wild-type litter mates. Transgenic mice expressing either growth hormone (GH) (PALMITER et al. 1982) or insulin-like growth factor (IGF) I (MATHEWS et al. 1988) grow larger than control mice (Fig. 2). During postpartum development from the fourth week onward, GH secreted from the pituitary gland in response to hypothalamic signals regulates postnatal body growth mainly by stimulating IGF I expression in peripheral tissues. IGF I subsequently stimulates mitogenesis in many tissues. Another growth factor, IGF II, is restricted to a growth-promoting role during embryonic development and growth in the first 3 weeks of postpartum development. The growth rate of $p27^{-/-}$ mice does not suggest that p27 gene disruption affects these endocrine pathways. About 70% of $p27^{-/-}$ mice are born larger than control littermates, and the rate of growth appears to approximate that observed in wild-type mice until the fourth or fifth week. At that time, the wild-type growth rate begins to plateau, but the $p27^{-/-}$ growth rate continues upward until it begins to plateau by the sixth week. Furthermore, the other 30% of the mice are born at the same weight as wild-type littermates, but by the fourth week they are larger. In addition, neither IGF I nor GH secretion is affected in p27-deficient mice. These observations imply that p27 disruption affects animal growth by acting within cells to alter the balance between proliferation and withdrawal from the cell cycle at critical periods of development rather than affecting the hormonal axis. Consistent with this hypothesis, embryonic fibroblasts and T lymphocytes isolated from p27-knockout mice are partially resistant to a growth-inhibitory agent, rapamycin (LUO et al. 1996). p27-Deficient keratinocytes are also partially refractory to transforming growth factor (TGF)-β-induced growth arrest (MISSERO et al. 1996). In addition, in thymocytes undergoing post natal maturation, there is an increase in the proportion of S-phase cells, and the differentiation of granulosa cells to luteal cells is impaired in $p27^{-/-}$ mice (see below). Thus p27 might be a critical molecule coordinating the cell cycle machinery and many developmental programs.

It is also important to note that, overall, the mouse embryonic fibroblasts isolated from $p27^{-/-}$ mice respond quite normally to anti-mitogenic conditions. High density or serum starvation leads to G_1-phase arrest in p27-deficient fibroblasts similar to that observed in controls. Furthermore, p27-deficient fibroblasts undergo senescence in a similar fashion to wild-type cells. These data suggest that either redundant proteins or more likely redundant mechanisms compensate for the loss of p27. Activation of CDK promotes S-phase entry, and many antimitogenic signals lead to a rapid increase in the p27 level and a slower reduction in the amount of cyclin or CDK protein. Consequently, loss of p27 simply prevents the fast reaction, but cells may arrest due to an intact slow reaction.

5.2 Thymic Development

Extensive studies to identify discrete developmental stages of thymocytes using cell surface markers have made it possible to study the details of cell cycle control during the multistep process of thymic development, such as TCR-β chain rearrangement, negative selection, and positive selection. Thymocytes in the process of β-chain rearrangement are arrested in the G_0/G_1 phase of the cell cycle, allowing recombination to occur without deleterious effects. Upon successful recombination, a signal (of unknown nature) is transduced and the cell enters S phase, undergoing proliferative expansion. Correlating with expansion are a decrease in the amount of p27, activation of CDK2 kinase, phosphorylation of the Rb gene product, and expression of CDC2 protein (HOFFMAN et al. 1996). The decrease in p27 suggests that this protein may act within a checkpoint coupling intracellular DNA rearrangement with cell proliferation.

The disruption of p27 in mice leads to an increase in the number of mature T lymphocytes, and there is an increase in thymocytes at virtually all stages of development leading to a markedly enlarged thymus. Hyperproliferation, and not a decrease in cell death, explains this phenotype. In situ bromodeoxyuridine (BrdU) labeling has shown that, in the p27-deficient thymus, there is an increase in the percentage of thymocytes in S phase. In contrast, apoptosis of thymocytes induced by irradiation or dexamethasone is not affected by p27 disruption (KIYOKAWA et al. 1996). These data suggest that the intracellular amount of p27 may determine the proportion of thymocytes in S phase during maturation.

5.3 Guardian of Quiescence

There is little evidence so far for genetic alterations of p27 in clinical tumors (BULLRICH et al. 1995; PIETENPOL et al. 1995; PONCE-CASTANEDA et al. 1995, KAWAMATA et al. 1995, HIRAMA and KOEFFLER 1995). In mice, p27 disruption results in nodular hyperplasia of the intermediate lobe in the pituitary gland with complete penetrance. This benign hyperplasia continues to enlarge as mice become older. These observations imply that in this lineage of cells p27 is indispensable for maintaining quiescence. It has yet to be determined whether malignant transformation occurs in the hyperplastic pituitary gland of old mice. This is of great interest, because loss of heterozygosity (LOH) in $Rb^{+/-}$ mice results in pituitary adenocarcinomas in the intermediate lobe (Hu et al. 1994). This suggests that p27 and Rb may be in the same genetic pathway or in closely interacting parallel pathways that maintain quiescence of these cells in the pituitary gland. It is consistent with this model that phosphorylation of Rb by G_1-phase CDK abrogates the growth-suppressive function of Rb (for a review, see WEINBERG 1995; also see Fig. 5). It should be noted that loss of Rb causes more aggressive tumorigenesis than loss of p27, implying that the pathway including p27 only partially regulates the function of Rb. At present it is not clear why p27 deficiency results in continuous hyperplasia of this specific cell type, while other cells in p27-knockout mice eventually stop proliferating following an increased proliferative period. Cells in the intermediate lobe may depend on p27 to establish quiescence, possibly because they lack other mechanisms to regulate G_1-phase CDK activity.

5.4 Female Fertility

Alteration in the decision of cells to proliferate could clearly affect alternative cell fates such as differentiation. Moreover, continual proliferation of cells destined to withdraw from the cell cycle might affect the entire organism by altering signaling between organs. Indeed, it is likely that female infertility of p27-deficient mice is due to a combination of impaired differentiation and defective signaling. NAKAYAMA et al. (1996) reported that the p27-deficient ovary did not develop mature graafian follicles, suggesting a defect in ovulation. In contrast, we have observed that p27-deficient females have spontaneous ovulation, although infrequently, and fertilized oocytes develop normally at least up to the morula stage (KIYOKAWA et al.

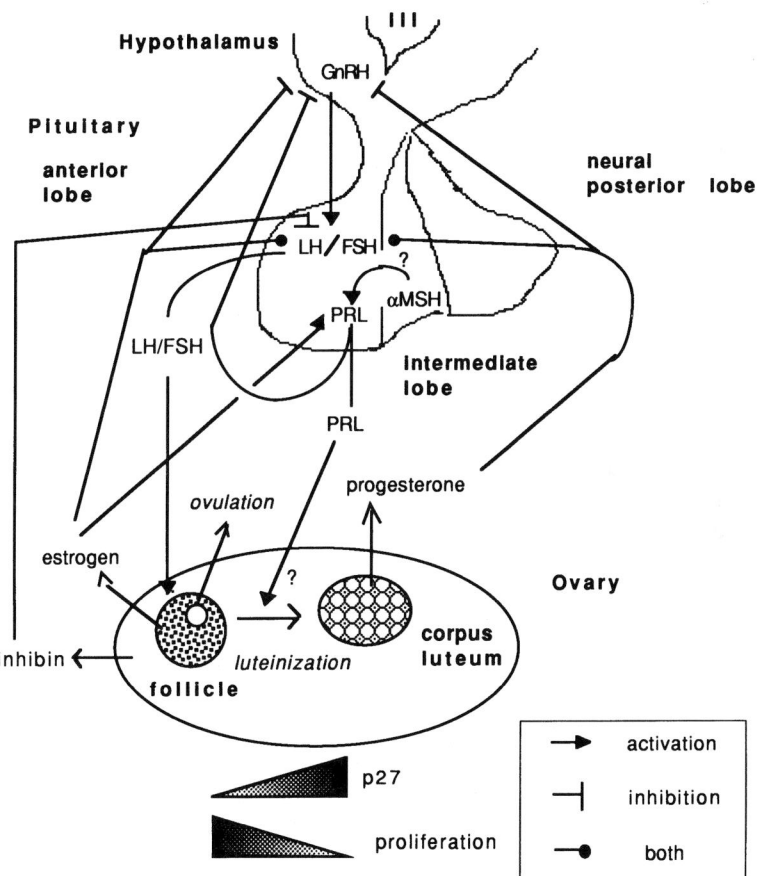

Fig. 3. Physiology of hormonal regulation in the hypothalamus–pituitary–ovary axis in mice. The amount of p27^{Kip1} and the proliferation activity during differentiation from granulosa cells into luteal cells are shown below the ovary scheme. *III*, third ventricum; *GnRH*, gonadotropin releasing hormone; *LH*, luteinizing hormone; *FSH*, follicle-stimulating hormone; *αMSH*, α-melanocyte-stimulating hormone; *PRL*, prolactin

1996). These different observations in the two laboratories might reflect a difference in the targeted mutations of p27 or in the genetic background of mice. We speculate that the infertility of p27-deficient females is, at least in part, due to their inability to maintain an appropriate maternal environment to allow full-term development of embryos. In the normal ovary, granulosa cells, the somatic component of the follicle, differentiate to luteal cells following ovulation. Concomitant with differentiation is the cessation of proliferation and induction of p27 protein expression (Fig. 3). In p27-deficient females, an intrinsic defect that prevents luteal cell differentiation may lead to a lack of corpus luteum function required for appropriate maternal environment. It is noteworthy that cyclin D_2-deficient female mice are sterile with hypoplastic ovarian follicles due to defective proliferation of granulosa cells (SICINSKI et al. 1996). Thus cyclin D_2 is required for proliferation of granulosa cells during follicle maturation, which is dependent on follicle-stimulating hormone. Induction of p27 following ovulation is critical for stopping proliferation in association with luteal differentiation. It is possible that, in granulosa cells, the switch to differentiation, which is triggered by luteinizing hormone, leads to suppression of dual pathways required for G_1S-phase transition. Decreased expression of cyclin D_2 may reduce the activity of cyclin D_2-dependent kinase, and induction of p27 expression may inhibit activity of cyclin E-dependent kinase. p27 may also participate in shutting off cyclin D_2-dependent kinase completely.

Altered signals between the pituitary gland and the ovary may also affect the formation of the corpus luteum. Physiological regulation in the hormonal axis among the hypothalamus, pituitary gland, and ovary is summarized in Fig. 3. The hyperplastic intermediate lobe may alter the pituitary function in p27-deficient mice. In addition, defective luteinization may affect the pituitary function, as corpus luteum formation is absolutely required to inform the pituitary gland that ovulation has occurred. The corpus luteum is maintained or disintegrated depending on whether oocytes are fertilized in the uterus. It is also possible that defects in the uterus related to implantation are partly, responsible for the sterility of p27-deficient females. These endocrine/paracrine signals are altered if cells in either the pituitary gland, the ovary, or the uterus do not respond appropriately. The primary defect leading to infertility has not yet been determined.

5.5 Retinal Development

An additional change in tissue organization as a result of p27 deficiency is observed in the retina. Some of the p27-deficient mice display marked disorganization in the layer formation of the retina (Fig. 4), although, unlike other phenotypes, the penetrance of this defect is incomplete. In the p27-deficient retina, the external nuclear layer, which contains the nuclei of the photoreceptor cells, protrudes into the layer of rods and cones, the cell bodies of the photoreceptor cells. Moreover, the cellularity of the internal nuclear layer, composed of the nuclei of bipolar cells, amacrine cells, and Müller cells, is also disturbed, making the outer plexiform layer irregular. Electroretinographic (ERG) analysis showed defective function of the

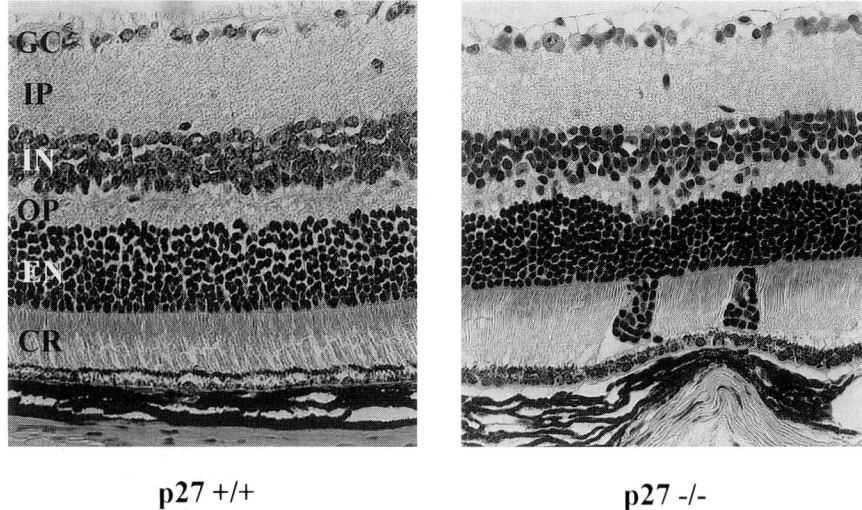

Fig. 4. Retinal dysplasia in an 8-week-old p27$^{Kip1-/-}$ mouse. *GC*, ganglion cell layer; *IP*, inner plexiform layer; *IN*, inner nuclear layer; *OP*, outer plexiform layer; *EN*, external nuclear layer; *CR*, cones and rods

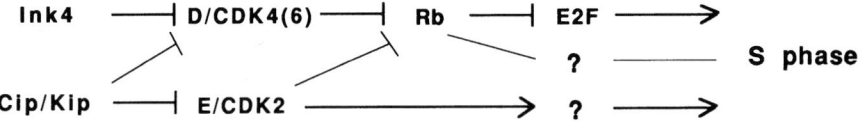

Fig. 5. Pathways leading to initiation of S phase of the cell cycle

photoreceptors in p27-deficient mice that had severe dysplasia of the retina (NAKAYAMA et al. 1996). Intriguingly, mice lacking cyclin D_1 have a hypoplastic retina with a marked decrease in cell numbers in all cell layers and show a severe reduction in ERG amplitude (SICINSKI et al. 1995; FANTL et al. 1995). In humans, loss of Rb function causes malignant transformation of the retinal cells, although for unknown reasons Rb deficiency does not cause retinoblastoma in mice (for a review, see WEINBERG 1995). Taken together, p27 may interact with the growth-regulatory pathway of retinal cells involving cyclin D_1 and Rb (Fig. 5). It remains to be elucidated whether p27-deficient mice will develop retinoblastomas at an older age. The changes in the external nuclear layer observed in the p27-deficient retina might reflect the clonal outgrowth of a subset of the photoreceptor cells.

6 p16^{Ink4A}, a Bona Fide Tumor Suppressor

It has been hypothesized that p16 has a tumor-suppressive function, since the 16 genomic locus is frequently inactivated in a variety of tumor-derived cell lines and clinical tumor samples by mutations, deletions, or methylation (for a review, see HIRAMA and KOEFFLER 1995). Furthermore, D-type cyclins, particularly cyclins D_1 and D_2, are overexpressed by gene rearrangement or amplification in many clinical cancers, and cyclin D-dependent kinases are the sole definite target of p16. Nonetheless, the tumor-suppressive function of p16 was officially proved by generation of p16-knockout mice. p16-mutant mice are viable without major defects in development, except for lighter coat pigmentation and possible alterations in hematopoiesis suggested by extramedullary hematopoiesis in the spleen. These mutants die with spontaneous malignant tumors, such as lymphomas and fibrosarcomas, at early ages comparable to p53-knockout mice (SERRANO et al. 1996). These mice are very sensitive to dimethylbenzanthracene (DMBA)- or ultraviolet (UV)-induced carcinogenesis. Moreover, fibroblasts obtained from these animals are prone to Ha-ras-induced transformation, implying that the loss of p16, like activation of myc, may collaborate with the activated ras pathway to induce malignant transformation. These observations indicate that p16 is a bona fide tumor suppressor. However, it is unclear why the spectrum of tumorigenesis in mice is different from that of clinical primary cancers with alterations of the p16 gene, i.e., melanoma, glioblastomas, esophageal cancers, pancreatic cancers, and acute lymphoblastic leukemia. This may reflect the fact that, in a large body of clinical cancers, deletion of p16 gene is accompanied by deletion of a closely related gene, $p15^{Ink4B}$, which is located only 25 kb from the p16 gene. This question awaits the generation and analysis of p15-knockout mice and a double mutant strain deficient for p16 and p15. In addition, the p16 genomic locus produces two different transcripts from different reading frames, p16 and $p19^{Arf}$ (QUELLE et al. 1995). Both p16 and $p19^{Arf}$ have growth-suppressive function in vitro and both are disrupted in p16-knockout mice. Thus the contribution of $p19^{Arf}$ deficiency to tumorigenesis in the p16-knockout mice remains to be clarified.

The knockout study also confirmed another role of p16 in senescence, a role originally suggested by the increased expression of p16 in cells undergoing senescence (HARA et al. 1996). Fibroblasts obtained from p16-deficient mice fail to undergo senescence after many passages. They also grow more rapidly and reach higher saturation density than wild-type cells, suggesting a role for p16 in cell cycle progression.

Ink4 proteins, including p16 and p15, specifically inhibit the cyclin D-dependent kinases CDK4 and CDK6. Phosphorylation of Rb by cyclin D-dependent kinases inactivates the growth-suppressive function of Rb, which is, at least in part, mediated by the transcription factor E2F (Fig. 5) Rb seems to be the most important substrate of cyclin D-dependent kinases, since cells lacking Rb neither require cyclin D for initiating S phase, nor do they arrest in G_1-phase in response to p16 overexpression. In contrast, Kip/Cip proteins inhibit both cyclin D- and cyclin E-dependent kinases in cells withdrawing from the cell cycle. These inhibitors can cause G_1-phase arrest irrespective of the Rb status, and cyclin E is required for S-phase initiation in all types of cells. These data suggest that the Kip/Cip–cyclin E pathway has one or more other substrates downstream which regulate the G_1/S-phase transition. It appears to be critical that essentially all the molecules on the pathways originating in Ink4 are directly involved in carcinogenesis. D-type cyclins are implicated in multiple types of cancers (for a review, see HIRAMA and KOEFFLER 1995), and Rb is of course a prototypical tumor suppressor. E2F-1 has been shown to transform cells in culture (for a review, see YAMASAKI et al. 1996). Paradoxically, E2F-1-knockout mice develop spontaneous tumors, including reproductive duct sarcomas, lung adenocarcinoma, and lymphomas (YAMASAKI, et al. 1996). This may be related to defective regulation of apoptosis, since E2F-1 has been shown to play a role in triggering apoptosis in vitro, and E2F-1-knockout mice exhibit defective apoptosis of thymocytes (FIELD et al. 1996).

7 Putative Physiological Roles of the CDK Inhibitors

None of the p21, p27 or p16 deficiency causes embryonic lethality, implying redundancy of the CDK inhibitor functions for developmental control of the cell cycle. p21-knockout mice have no developmental phenotypes, but cells derived from these animals display defective G1 checkpoint function in response to DNA damage and nucleotide pool perturbation. p27-knockout mice show marked de-

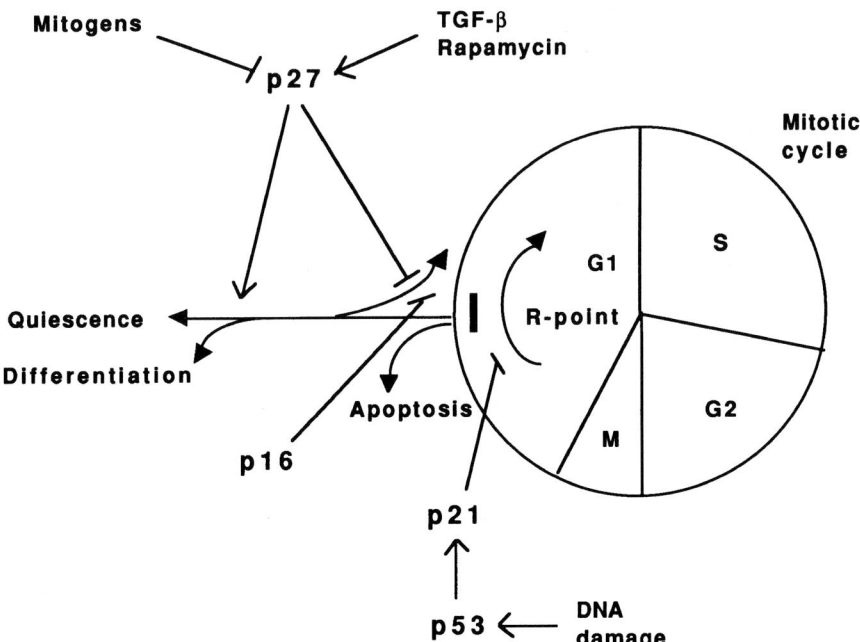

Fig. 6. Putative roles for the cyclin-dependent kinase inhibitors p21^{Cip1}, p27^{Kip1}, and p16^{Ink4A} in the mammalian cell cycle. The restriction point in G_1 phase is shown as *R-point*. *TGF*, transforming growth factor

velopmental phenotypes, such as enhanced growth, female infertility, and retinal dysplasia, while these animals also develop nodular hyperplasia in the pituitary. These observations suggest a putative role for p21 as a regulator of CDK activity once cells have committed to the mitotic cycle and for p27 to regulate CDK activity before commitment to the mitotic cycle (Fig. 6). Thus, p27 could act as a rate-limiting regulator for exit from the mitotic cycle and p21 as an SOS-like response to prevent replication once cells have committed to the cycle. Further support for this hypothesis comes from studies on tissue culture cells. During mitogen-stimulated entry of T-cells into the cell cycle, the level of p21 increases and p27 decreases (NOURSE et al. 1995). In A431 cells deprived of EGF cells arrest by a p27-dependent mechanism; however, arrest due to hyperactivation of the EGF receptor occurs by a p21-dependent mechanism (Fan et al, unpublished data). Thus the studies of the Cip/Kip family of CDK inhibitors in mice suggest that although these two proteins have similar biochemical functions their physiological roles are not necessarily redundant. Proof of this hypothesis awaits the analysis of double and triple mutants deficient for p21, p27 and p57. In contrast, the knockout study of p16 has confirmed that p16 is a bona fide tumor suppressor protein. p16, D-type cyclins, Rb, and E2F seem to function on the same genetic pathway directly involved in carcinogenesis.

Acknowledgements. This work was supported in part by grants from the National Institute of Health (GM52597) and the Society of Memorial Sloan-Kettering Cancer Center. H.K. is supported by the DeWitt Wallace Research Fund of the Memorial Sloan-Kettering Cancer Center. A.K. is a Pew Scholar in Biomedical Science and the recipient of the Frederick R. Adler Chair for the Junior Faculty.

References

Abbondanzo SJ, Gadi I, Stewart CL (1993) Derivation of embryonic stem cell lines. Methods Enzymol 225:803–823
Bhatia K, Fan S, Spangler G, Weintraub M, O'Connor PM, Judde J-G, Magrath I (1995) A mutant p21 cyclin-dependent kinase inhibitor isolated from a Burkitt's lymphoma. Cancer Res 55:1431–1435
Brugarolas J, Chandrasekaran C, Gordon JI, Beach D, Jacks T, Hannon GJ (1995) Radiation-induced cell cycle arrest compromised by p21 deficiency. Nature 377:552–557
Bullrich F, MacLachlan TK, Sang N, Druck T, Veronese ML, Allen SL, Chiorazzi N, Koff A, Heubner K, Croce CM, Giordano A (1995) Chromosomal mapping of members of the cdc2 family of protein kinases, cdk3, cdk6, PISSLRE, and PITALRE, and a cdk inhibitor, p27^{Kip1}, to regions involved in human cancer. Cancer Res 55:1199–1205
Cross SM, Sanchez CA, Morgan CA, Schimke MK, Ramel S, Idzerda RL, Raskind WH, Reid BJ (1995) A p53-dependent mouse spindle checkpoint. Science 267:1353–1356
Deng C, Zhang P, Harper JW, Elledge SJ, Leader P (1995) Mice lacking p21$^{CIP1/WAF1}$ undergo normal development, but are defective in G1 checkpoint control. Cell 82:675–684
Donehower LA, Harvey M, Slagle BL, McArthur MJ, Montgomery CA, Butel JS, Bradley A (1992) Mice deficient for p53 are developmentally normal but susceptible to spontaneous tumours. Nature 356:215–221
Dulic V, Kaufmann WK, Wilson SJ, Tlsty TD, Lees E, Harper JW, Elledge SJ, Reed SI (1994) p53-dependent inhibition of cyclin-dependent kinase activities in human fibroblasts during radiation-induced G1 arrest. Cell 76:1013–1023
El-Deiry WS, Tokino T, Velculescu VE, Levy DB, Parsons R, Trent JM, Lin D, Mercer E, Kinzler KW, Vogelstein B (1993) WAF1, a potential mediator of p53 tumor suppression. Cell 75:817–825

Fantl V, Stamp G, Andrews A, Rosewell I, Dickson C (1995) Mice lacking cyclin D1 are small and show defects in eye and mammary gland development. Genes Dev 9:2364–2372

Fero MI, Rivkin M, Tasch M, Porter P, Carow CE, Firpo E, Polyak K, Tsai L-H, Broudy V, Perlmutter RM, Kaushansky K, Roberts JM (1996) A syndrome of multi-organ hyperplasia with features of gigantism, tumorigenesis and female sterility in p27Kip1-deficient mice. Cell 85:733–744

Field SJ, Tsai F-Y, Kuo F, Zubiaga AM, Kaelin WG, Livingston DM, Orkin SH, Greenberg ME (1996) E2F-1 functions in mice to promote apoptosis and suppress proliferation. Cell 85:549–561

Halevy O, Novitch BG, Spicer DB, Skapek SX, Rhee J, Hannon GJ, Beach D, Lassar AB (1995) Correlation of terminal cell cycle arrest of skeletal muscle with induction of p21 by MyoD. Science 267:1018–1021

Hara E, Smith R, Parry D, Tahara H, Stone S, Peters G (1996) Regulation of $p16^{CDKN2}$ expression and its implications for cell immortalization and senescence. Mol Cell Biol 16:859–867

Hirama T, Koeffler HP (1995) Roles of the cyclin-dependent kinase inhibitors in the development of cancer. Blood 86:841–854

Hoffman ES, Passoni L, Crompton T, Leu TMJ, Schatz DG, Koff A, Owen MJ, Hayday AC (1996) Productive T-cell receptor β-chain gene rearrangement: coincident regulation of cell cycle and clonality during development in vivo. Genes Dev 10:948–962

Hu N, Gutsmann A, Herbert DC, Bradley A, Lee W-H, Lee EY-HP (1994) Heterozygous $Rb-1^{20/+}$ mice are predisposed to tumors of the pituitary gland with a nearly complete penetrance. Oncogene 9:1021–1027

Kawamata N, Morosetti R, Miller CW, Park D, Spirin KS, Nakamaki T, Takeuchi S, Hatta Y, Simpson J, Wilczynski S, Lee YY, Bartram CR, Koeffler HP (1995) Molecular analysis of the cyclin-dependent kinase inhibitor gene p27/Kip1 in human malignancies. Cancer Res 55:2266–2269

Kiyokawa H, Kineman RD, Manova-Todorova KO, Soares VC, Hoffman E, Ono M, Khanam D, Hayday AC, Frohman LA, Koff A (1996) Enhanced growth of mice lacking the cyclin-dependent kinase inhibito function of $p27^{Kip1}$. Cell 85:721–732

Koff A, Polyak K (1995) $p27^{KIP1}$, an inhibitor of cyclin-dependent kinases. In: Meijer L, Guidet S, Tung HYL (ed) Progress in cell cycle research, vol.1. Plenum, New York, pp 141–147

Kranenburg O, Scharnhorst V, Van der Eb A, Zantema A (1995) Inhibition of cyclin-dependent kinase activity triggers neuronal differentiation of mouse neuroblastoma cells. J Cell Biol 131:227–234

Liu M, Lee MH, Cohen M, Bommakanti M, Freedman LP (1996) Transcriptional activation of the Cdk inhibitor p21 by vitamin D3 leads to the induced differentiation of the myelomonocytic cell line U937. Genes Dev 10:142–153

Luo Y, Marx SO, Kiyokawa H, Koff A, Massague J, Marks A (1996) Rapamycin resistance tied to defective regulation of $p27^{Kip1}$. Mol Cell Biol 16:6744–6751

Mansour SL, Thomas KR, Capecchi MR (1988) Diruption of the proto-oncogene int-2 in mouse embryo-derived stem cells: a general strategy for targeting mutations to non-selectable genes. Nature 336:348–352

Mathews LS, Hammer RE, Behringer RR, D'ercole AJ, Bell GI, Brinster RL, Palmiter RD (1988) Growth enhancement of transgenic mice expressing human insulin-like growth factor I. Endocrinology 123:2827–2833

Matsuoka S, Edwards MC, Bai C, Parker S, Zhang P, Baldini A, Harper JW, Elledge SJ (1995) $p57^{KIP2}$, a structurally distinct member of the $p21^{CIP1}$ Cdk inhibitor family, is a candidate tumor suppressor gene. Genes Dev 9:650–662

Missero C, Di Cunto F, Kiyokawa H, Koff A, Dotto GP (1996) The absence of p21Cip1/WAF1 alters keratinocyte growth and differentiation and promotes ras-tumor progression. Genes Dev 10:3065–3075

Nakayama K, Ishida N, Shirane M, Inomata A, Inoue T, Shishido N, Horii I, Loh DY, Nakayama K-I (1996) Mice lacking $p27^{Kip1}$ display increased body size, multiple organ hyperplasia, retinal dysplasia, and pituitary tumors. Cell 85:707–720

Nourse J, Firpo E, Flanagan WM, Coats S, Polyak K, Lee M-H, Massague J, Crabtree GR, Roberts JM (1994) Interleukin-2-mediated elimination of p27kip1 cyclin-dependent kinase inhibitor prevented by rapamycin. Nature 372:570–573

Palmiter RD, Brinster RL, Hammer RE, Trumbauer ME, Rosenfeld MG, Brinberg NC, Evans RM (1982) Dramatic growth of mice that develop from eggs microinjected with metalothionein-growth hormone fusion genes. Nature 300:611–615

Parker SB, Eichele G, Zhang P, Rawls A, Sands AT, Bradley A, Olson EN, Harper JW, Elledge SJ (1995) p53-independent expression of $p21^{Cip1}$ in muscle and other terminally differentiating cells. Science 267:1024–1027

Pietenpol JA, Bohlander SK, Sato Y, Papadopoulos N, Liu B, Friedman C, Trask BJ, Roberts JM, Kinzler KW, Rowley JD, Vogelstein B (1995) Assignment of the human p27^{Kip1} gene to 12p13 and its analysis in leukemias. Cancer Res 55:1206–1210

Ponce-Castaneda MV, Lee M-H, Latres E, Polyak K, Lacombe L, Montgomery K, Mathew S, Krauter K, Sheinfeld J, Massague J, Cordon-Cardo C (1995) p27^{Kip1}: chromosomal mapping to 12p12–12p13.1 and absence of mutations in human tumors. Cancer Res 55:1211–1214

Quelle DE, Zindy F, Ashmun RA, Sherr CJ (1995) Alternative reading frames of the INK4a tumor suppressor gene encode two unrelated proteins capable of inducing cell cycle arrest. Cell 83:993–1000

Ramirez-Solis R, Davis AC, Bradley A (1993) Gene targeting in embryonic stem cells. Methods Enzymol 225:855–878

Serrano M, Lee H-W, Chin L, Cordon-Cardo C, Beach D, DePinho RA (1996) Role of the INK4a locus in tumor suppression and cell mortality. Cell 85:27–37

Sherr CJ, Roberts JM (1995) Inhibitors of mammalian G1 cyclin-dependent kinases. Genes Dev 9:1149–1163

Shiohara ML, El-Deiry WS, Wada M, Nakamaki T, Takeuchi S, Yang R, Chen DL, Vogelstein B, Koeffler HP (1994) Absence of WAF1 mutations in a variety of human malignancies. Blood 84:3781–3784

Sicinski P, Donaher JL, Parker SB, Li T, Fazeli A, Gardner H, Haslam SZ, Bronson RT, Elledge SJ, Weinberg RA (1995) Cyclin D1 provides a link between development and oncogenesis in the retina and breast. Cell 82:621–630

Sicinski P, Donaher JL, Geng Y, Parker SB, Gardner H, Park MY, Robker RL, Richards JS, McGinnis LK, Biggers JD, Eppig JJ, Bronson RT, Elledge SJ, Weinberg RA (1996) Cyclin D2 is an FSH-responsive gene involved in gonadal cell proliferation and oncogenesis. Nature 384:470–474

Skapek SX, Rhee J, Spicer DB, Lassar AB (1995) Inhibition of myogenic differentiation in proliferating myoblasts by cyclin D1-dependent kinase. Science 267:1022–1024

Soos TJ, Kiyokawa H, Yan JS, Rubin MS, Giordano A, DeBlasio A, Bottega S, Wong B, Mendelsohn J, Koff A (1996) Formation of p27-CDK complexes during human mitotic cell cycle. Cell Growth Differ 7:135–146

Weinberg RA (1995) The retinoblastoma protein and cell cycle control. Cell 81:323–330

Yamasaki L, Jacks T, Bronson R, Goillot E, Harlow E, Dyson NJ (1996) Tumor induction and tissue atrophy in mice lacking E2F-1. Cell 85:537–548

p21/p53, Cellular Growth Control and Genomic Integrity

W.S. El-Deiry

1 Introduction 121
2 The Structure of p21 122
3 The Function of p21 122
4 The Regulation of p21 125
5 p53-Dependent Growth Inhibition and Genetic Fidelity 127
6 p53/p21 and Genomic Instability 130
References 132

1 Introduction

In the short time since its discovery in 1993, the prototype cyclin-dependent kinase (CDK) inhibitor p21 (CIP1, Harper et al. 1993; WAF1, El-Deiry et al. 1993; p21, Xiong et al. 1993a; CAP20, Gu et al. 1993; SDI1, Noda et al. 1994; MDA6, Jiang et al. 1995) has become an intensely studied molecule. Much has been learned about its molecular structure, biochemical interactions, and physiological role. p21 is a focal point that integrates many types of signals that impact on processes of cell division and cell death. Since 1995 to date there have been over 600 references in Current Contents dealing with WAF1 or CIP1 and they are increasing at a rate of up to 25 per week. The p53 field which dates back to 1979 has had some 6000 references on Medline since 1993. This chapter will highlight some of the recent discoveries emphasizing the role of p21 and p53 in growth control and the maintenance of genomic integrity.

Laboratory of Molecular Oncology and Cell Cycle Regulation, Howard Hughes Medical Institute, Department of Medicine, Genetics and Cancer Center, University of Pennsylvania School of Medicine, Philadelphia, PA 19104, USA

2 The Structure of p21

The WAF1/CIP1 gene is composed of three exons located within a 10-kb region on human chromosome 6p21.2 (EL-DEIRY et al. 1993). The 2.1-kb transcript encodes a conserved 164-amino acid 21-kD polypeptide which localizes to the nucleus of mammalian cells (EL-DEIRY et al. 1993, 1994). About two thirds of the human p21 transcript is 3'-untranslated sequence. Homologues of p21 have been cloned from mouse (EL-DEIRY et al. 1993; HUPPI et al. 1994), rat (EL-DEIRY et al. 1995), and cat (OKUDA et al. 1997). Homologues of p21 are likely to exist in *Xenopus* (SHOU and DUNPHY 1996; COX et al. 1994), insects (BAE et al. 1995), drosophila (DE NOOIJ et al. 1996), plants (BALL and LANE 1996), and yeast (ELLEDGE 1996).

It was recognized early on that p21 is part of a family of universal cyclin-CDK inhibitors and that this is mediated by an amino-terminal domain which is homologous among p21, p27, and p57 (for review see EL-DEIRY 1996). It was subsequently found that the cyclin-CDK inhibitory domain of p21 contains distinct cyclin- and CDK-interacting regions (LIN et al. 1996). The crystal structure of cyclin A-CDK2-p27 (N-terminal cyclin-CDK inhibitory domain) provides insight into how these small kinase inhibitory molecules bind cyclin-CDK complexes and arrest cell growth (RUSSO et al. 1996). The p27 bound cyclin-CDK complex has an altered kinase conformation and there is steric interference with ATP binding at the active site. Although it is likely that p21 may similarly affect CDK structure, it would not be surprising if there were important differences given the existence of important biochemical and biological differences (LABAER et al. 1997; MISSERO et al. 1996; REYNISDOTTIR and MASSAGUE 1997).

One of the early observations was that cyclin-CDK subunits became rearranged in transformed cells (XIONG et al. 1993b). In particular, it was found that normal cells contain p21 and proliferating-cell nuclear antigen (PCNA) associated with various cyclin-CDK complexes. In fact, it was later shown that p21 is a potent inhibitor of processive DNA synthesis through its interaction with PCNA (WAGA et al. 1994; LI et al. 1994; CHEN et al. 1995b; LUO et al. 1995). This PCNA-interacting domain resides in the carboxy-terminal region of p21, which also contains its nuclear localization signal. The crystal structure of a p21-PCNA-DNA complex was recently reported (GULBIS et al. 1996). The structure suggests that p21 probably masks contact sites on PCNA that may permit association and function of other replication proteins (GULBIS et al. 1996). There has been recent biochemical evidence to support this possibility (CHEN et al. 1996).

3 The Function of p21

p21 is not required for normal mouse development (BRUGAROLAS et al. 1995; DENG et al. 1995), although it may be required in drosophila (DE NOOIJ et al. 1996; LANE et al. 1996). p21-null mice have no increased predisposition to tumor development (BRUGAROLAS et al. 1995; DENG et al. 1995), at least in the absence of carcinogens. p21 is required for the DNA damage- and other p53-dependent checkpoint controls (Table 1). In p21−/− mouse cells there is a partial deficiency in G1 cell cycle arrest (but not in apoptosis induction) following radiation or nucleotide depletion (BRUGAROLAS et al. 1995; DENG et al. 1995). In p21−/− human colon cancer cells a complete loss of G1 cell cycle arrest was reported, and this was associated with a predisposition to S- and M-phase uncoupling following exposure to DNA-damaging drugs (WALDMAN et al. 1995, 1996). An altered sensitivity of p21−/− cells to certain DNA-damaging chemotherapeutic drugs has been demonstrated both in vitro (MCDONALD et al. 1996; WALDMAN et al. 1996; LI et al. 1997; FAN et al. 1997) and in vivo (WALDMAN et al. 1997). p21 −/− cells have been shown to be defective in nucleotide excision repair and it has been suggested that this may underlie their increased sensitivity to certain chemotherapeutic drugs (MCDONALD et al. 1996).

Besides its role in p53-dependent checkpoint control of cell-cycle progression, p21 has been found to be involved in mediating a p53-independent growth arrest associated with terminal differentiation. Although this is clearly not required for mouse development, it is likely that p21 plays a role in inducing a quiescent state which is permissive for differentiation. There is some evidence that p21 expression alone may induce differentiation (LIU et al. 1996a; SHEIKH et al. 1995; MENG et al.

Table 1. Functions performed by p21

Functions attributed to p21$^{WAF1/CIP1}$	Reference
p53-Dependent growth suppression	EL-DEIRY et al. 1993
Inhibition of cylin-dependent kinase activity	XIONG et al. 1993a, b; HARPER et al. 1993, 1995
Inhibition of CDK phosphorylation by CAK	APRELIKOVA et al. 1995
Cell cycle arrest after DNA damage	DULIC et al. 1994; EL-DEIRY et al. 1994; DENG et al. 1995; BRUGAROLAS et al. 1995; WALDMAN et al. 1995
Differentiation-associated growth arrest	ZHANG et al. 1995; PARKER et al. 1995; EL-DEIRY et al. 1995; ZENG et al. 1997
Senescence-associated growth arrest	NODA et al. 1994; SERRANO et al. 1997
Shift from replication to repair	LI et al. 1994; LUO et al. 1995; CHEN et al. 1996; MCDONALD et al. 1996; SHEIKH et al. 1997
Inhibition of SAP kinase activity	SHIM et al. 1996
Regulation of NF-κB activity	PERKINS et al. 1997
Inhibition of transformation	GIVOL et al. 1995; MICHIELI et al. 1996; MISSERO et al. 1996; SOMASUNDARAM and EL-DEIRY 1997
Response to ECM, cytoskeleton	BOHMER et al. 1996; SANTRA et al. 1997
Inhibition of apoptosis	GOROSPE et al. 1996; WANG and WALSH 1996; POLYAK et al. 1996; WALDMAN et al. 1996, 1997
Induction of apoptosis/giant cells	SHEIKH et al. 1995; PRABHU et al. 1996; LI et al. 1997; MENG et al. 1997; SHEIKH et al. 1997; FAN et al. 1997
Coupling S- and M-phase of the cell cycle	WALDMAN et al. 1996
Cyclin-CDK assembly and activation	ZHANG et al. 1994; LABAER et al. 1997

CDK, cyclin-dependent kinase; CAK, CDK-activating kinase; SAP, stress-activated protein kinase; NF-κB, nuclear factor-κB; ECM, extracellular matrix.

1997) and that in a p21-null background there is loss of differentiation marker expression in vivo (MISSERO et al. 1996). There is also evidence that p21 expression is increased in some fully differentiated tissues, such as muscle or gastrointestinal epithelium (PARKER et al. 1995; EL-DEIRY et al. 1995), suggesting the possibility that p21 may help maintain their growth arrest.

p21 is a potent inhibitor of cell growth both in vitro (EL-DEIRY et al. 1993) and in vivo (Wu et al. 1996). The growth inhibitory property of p21 has been mapped to its amino-terminal cyclin-CDK inhibitory domain (ZAKUT and GIVOL 1995; PRABHU et al. 1996). Although it could suppress colony formation, wild-type p21 overexpression does not eliminate colony formation, as has been observed for p53 (EL-DEIRY et al. 1993; PRABHU et al. 1996). Why some cancers escape growth inhibition despite p21 expression remains unexplained. It is clear that the PCNA-interacting domain of p21 is not sufficient for cancer cell growth inhibition, in the absence of the cyclin-CDK inhibitory domain (LUO et al. 1995). The cyclin-CDK inhibitory domain was found to be a more potent cancer growth suppressor compared to wild-type p21 (PRABHU et al. 1996).

The growth inhibitory effect of p21 has been found to suppress cellular transformation (GIVOL et al. 1995; MICHIELI et al. 1996). In fact, it was recently shown that p-21 deficient mouse keratinocytes are highly predisposed to *ras*-induced transformation and the development of highly malignant tumors in mice (MISSERO et al. 1996). p21 expression is generally decreased in tumors, primarily through p53 dysfunction (EL-DEIRY et al. 1994, 1995). Although very rare (SHIOHARA et al. 1994), mutations in p21 have been reported in human tumors (BALBIN et al. 1996; MALCOWICZ et al. 1996). Some of these tumor-derived mutant p21 proteins have been found to have loss of function (BALBIN et al. 1996). There appears to be a somewhat higher frequency of p21 alterations in bladder and prostate cancer, for unknown reasons (MALKOWICZ et al. 1996). Adenovirus E1A has been found to inhibit p21 expression through its p300/CBP-interacting domain and it has been hypothesized that such loss of p21 expression may contribute to E1A-mediated transformation (SOMASUNDARAM and EL-DEIRY 1997).

There are additional functions or roles that have been suggested for p21 (Table 1 and references therein). These include a role in senescence, apoptosis, stress-activated protein kinase regulation, and a docking function that may facilitate cyclin-CDK assembly and possibly their recruitment to replication centers. The role of p21 in apoptosis remains somewhat controversial. Experiments have suggested that p21 may protect cells against DNA damage or prostaglandin-induced apoptosis (GOROSPE et al. 1996; WANG and WALSH 1996; POLYAK et al. 1996; WALDMAN et al. 1996, 1997). There is also evidence that p21 is expressed in p53-dependent apoptosis (EL-DEIRY et al. 1994) and that its overexpression may contribute to apoptosis (SHEIKH et al. 1995, 1997; PRABHU et al. 1996; LI et al. 1997; MENG et al. 1997; FAN et al. 1997). It was recently observed that the primary effect of p21 overexpression is induction of a large cell-quiescent phenotype associated with a less prominent, although significant, increase in DNA ladder formation (SHEIKH et al. 1995, 1997; PRABHU et al. 1996; MENG et al. 1997).

Thus, p21 may assist in cyclin-CDK assembly, activation of kinase activity at low stoichiometry, inhibition of cyclin-CDK dissociation, and ultimately at higher stoichiometry, inhibition of cyclin-dependent kinase activity (ZHANG et al. 1994; LaBAER et al. 1997). A role in recruitment of kinases to replication centers has been proposed (ZHU et al. 1995). Evidence for p21-dependent inhibition of CDK phosphorylation by CDK-activating kinase (CAK) has been found (APRELIKOVA et al. 1995). Nonetheless, the major critical function of p21 appears to be inhibition of cyclin-CDK activity, an effect that causes cell cycle arrest and inhibition of cell division. The strongest evidence for this comes from the p21 knockout mouse fibroblasts (DENG et al. 1995; BRUGAROLAS et al. 1995) and p21 knockout human colon cancer cells (WALDMAN et al. 1995), both of which have deficient cell cycle arrest following DNA damage. The growth arrest associated with inhibition of cyclin-CDK activity probably provides an opportunity for DNA repair. p21 is involved in such repair through its interaction with PCNA (WAGA et al. 1994). Through this interaction, p21 inhibits processive DNA synthesis (LI et al. 1994), inhibits PCNA association with FEN1, thereby inhibiting 5′–3′ exonucleolytic removal of RNA primers during DNA synthesis (CHEN et al. 1996), and stimulates nucleotide excision repair (McDONALD et al. 1996; SHEIKH et al. 1997). Through its effect on repair, as well as maintaining the coordination of S-phase and M-phase synchrony (WALDMAN et al. 1996), p21 protects mammalian cells from death. The growth arrest induced by p21 may be required for differentiation of certain cell types and may be required to maintain the differentiation of certain cells (Fig. 1; Table 1 and references therein). p21 expression may suppress transformation and when overexpressed, in addition to quiescence, some cells may undergo apoptosis. It is believed that the effects of p21 on cell growth help to maintain genomic integrity.

4 The Regulation of p21

Initial observations indicated that p53 regulates basal as well as DNA damage-inducible p21 expression (EL-DEIRY et al. 1993, 1994). The vast majority of cells containing functional p53 (in terms of checkpoint control) express higher basal levels of p21 as compared to cells containing mutant p53 or are otherwise deficient in p53 expression or function (EL-DEIRY et al. 1994, 1995). Transcriptional activation of p21 expression is not observed following exposure of p53-deficient cells to DNA-damaging agents such as γ-radiation or (low doses of) topoisomerase II inhibitors (MICHIELI et al. 1994; EL-DEIRY et al. 1994). High doses of adriamycin (doxorubicin), a useful chemotherapeutic agent in the clinic, have been shown to increase p21 expression in a p53-independent manner (MICHIELI et al. 1994). It is clear that adriamycin, whose primary effect on the mammalian cell cycle is a G2-phase arrest (WALDMAN et al. 1995, 1996), is cytotoxic to cells through p53-dependent, as well as p53-independent, mechanisms. However, there is little evidence to suggest that p21

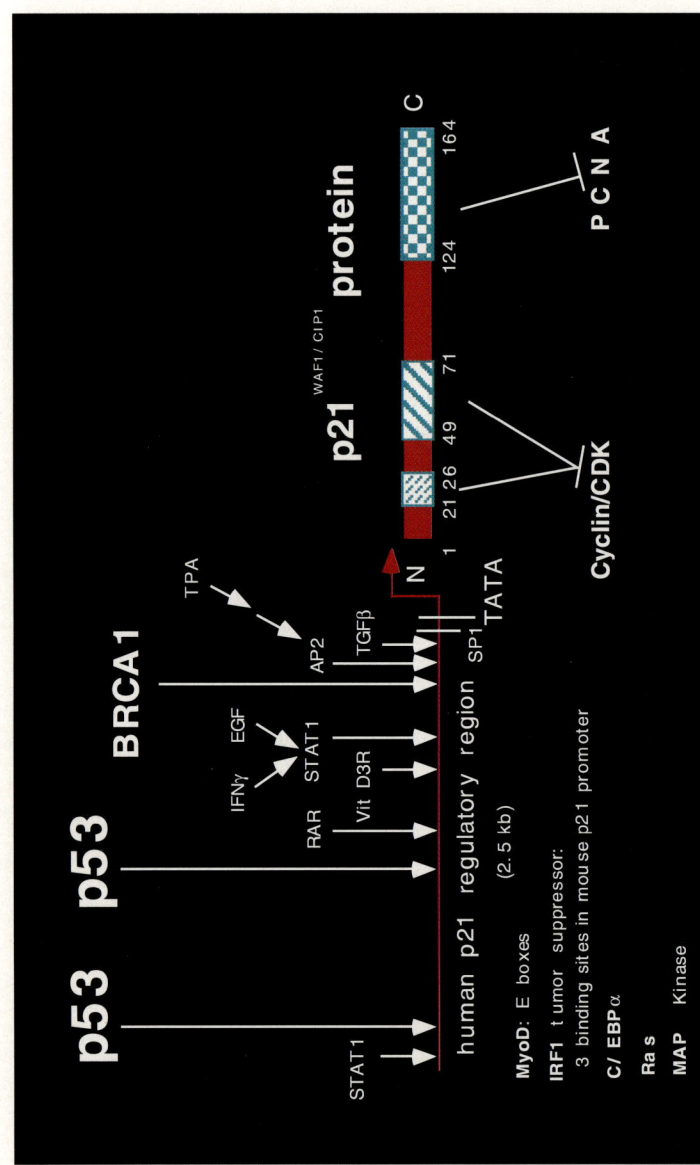

Fig. 1. Regulation of p21(WAF1/CIP1) expression by tumor suppressor genes, growth promoting factors, and differentiation associated transcription factors. More detailed information on the regulators of p21 is available in the original references (p53, EL-DEIRY et al. 1993, 1995; BRCA1, SOMASUNDARAM et al. 1997; IRF1, TANAKA et al. 1996; MyoD, PARKER et al. 1995; AP2, ZENG et al. 1997; Vitamin D3 receptor and RAR, LIU et al. 1996a,b; C/EBPα, TIMCHENKO et al. 1996; IFNγ, EGF, and STAT1, CHIN et al. 1996; *ras*, LIU et al. 1996d; SERRANO et al. 1997; MAP kinase, LIU et al. 1996d; TGFβ, DATTO et al. 1995). (Mapping of the separate cyclin and cyclin-dependent kinase, *CDK*, binding domains was obtained from LIN et al. 1996). *PCNA*, Proliferating-cell nuclear antigen

is involved in G2 arrest. In fact, p21 binds weakly to cyclin B/cdc2 complexes as compared to G1-phase cyclin-CDKs (HARPER et al. 1995), and p21−/− cells arrest in G2 following exposure to DNA-damaging agents (WALDMAN et al. 1995, 1996). It appears that S-phase arrest following DNA damage may also be p21-independent (WYLLIE et al. 1996).

It was quickly learned that regulation of p21 expression is complex, involving transcriptional and post-transcriptional mechanisms. Serum, a number of growth factors, an increasing number of cytotoxic and differentiating agents augment p21 expression through p53-independent pathways (MICHIELI et al. 1994). Some of the transcription factors have been identified and mapped within the p21 promoter (Fig. 1 and references therein). Evidence also suggests that p21 mRNA and protein stability vary under certain conditions, consistent with the possibility that they may affect its physiological function (LI et al. 1996; BLAGOSKLONNY et al. 1996). Alternative views of p21 activation by growth factors have been proposed. One view is that at some expression levels p21 may activate cyclin-CDK complexes (ZHANG et al. 1994; LABAER et al. 1997). There is evidence to support the notion that, depending on the stoichiometry between p21 and cyclin-CDK complexes, kinase activity may be stimulated or inhibited (ZHANG et al. 1994; LABAER et al. 1997). However, there does not appear to be a deficiency in kinase assembly or activity in a p21-null background (LABAER et al. 1997). It has also been proposed that p21 activation by growth factors may delay progression into S-phase, as a checkpoint that probably functions through Rb and restriction point control (ZENG and EL-DEIRY 1996). The extracellular matrix–integrin signaling pathway and specific proteoglycans have been found to regulate p21 expression (ZHU et al. 1996; SANTRA et al. 1997). It is clear that a growing list of differentiation-associated transcription factors activate p21 expression, leading to growth arrest that is probably necessary for differentiation (Fig. 1 and references therein). Little is known about the control of p21 in senescence, but recently it was shown that *ras* can increase p53 and p21 expression in its acceleration of senescence (SERRANO et al. 1997).

A wide variety of extracellular and intracellular signals therefore regulate p21 expression/function through transcriptional and post-transcriptional mechanisms. p21 serves as a focal point linking these growth suppressive/differentiation signals to cell cycle control. Loss of p21 expression, or its normal regulation, is associated with abnormal checkpoint control, abnormal differentiation, and a strong predisposition towards cellular transformation by oncogenes such as *ras*. Abnormalities in checkpoint control have been associated with increased mutation rates, gene amplification, and genomic instability (see below).

5 p53-Dependent Growth Inhibition and Genetic Fidelity

Multiple mechanisms ensure that the genetic material of cells is faithfully replicated and preserved. These mechanisms include thermodynamic constraints imposed by

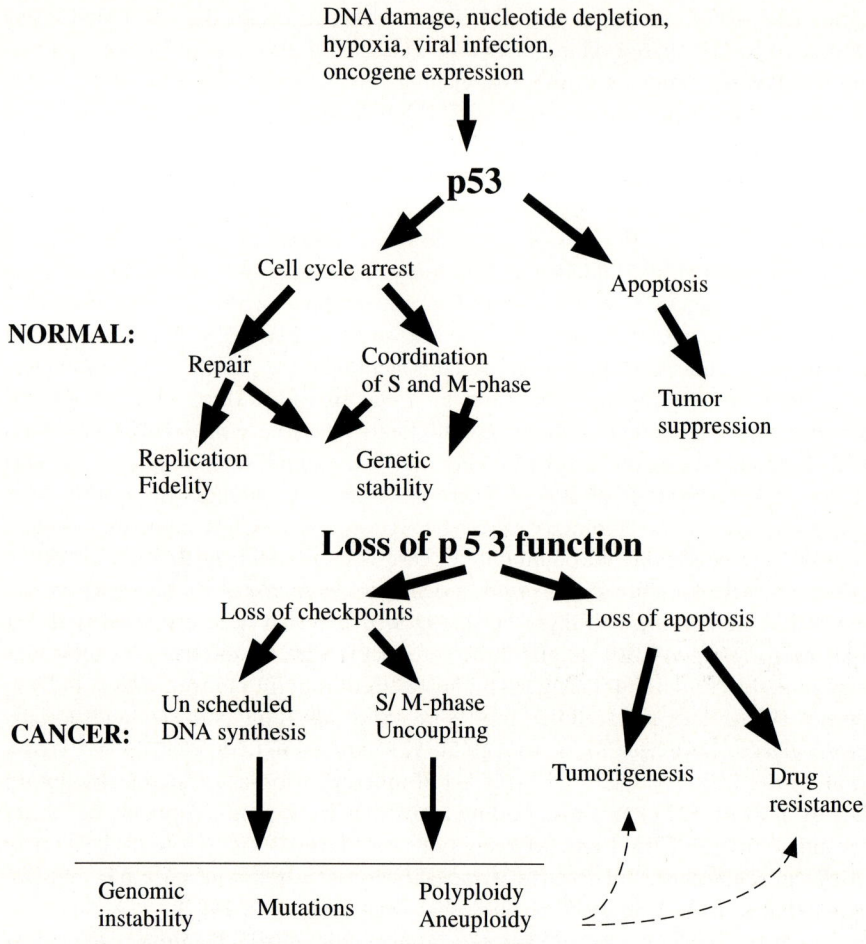

Fig. 2. p53-Dependent activation of pathways leading to tumor suppression and genomic stability and the consequences of p53 loss or mutation in cancer cells

the structure of DNA, selectivity by the DNA polymerases, replication-associated proofreading, and post-replication associated error or damage repair. Failure of these processes can result in mutagenesis and may lead to cancer development in higher organisms. Interference with these processes is useful in anti-viral and anti-cancer therapy. Certain signals such as nucleotide depletion, DNA damage, hypoxia or viral infection, or oncogene activation can trigger cell cycle checkpoints which are designed to ensure the integrity of a cell's genetic material. The p53 tumor suppressor protein is a critical regulator of this checkpoint control (Figs. 2 and 3).

The p53 tumor suppressor gene is also the most common target for alteration in human cancer (for reviews see LEVINE 1997; VELCULESCU and EL-DEIRY 1996).

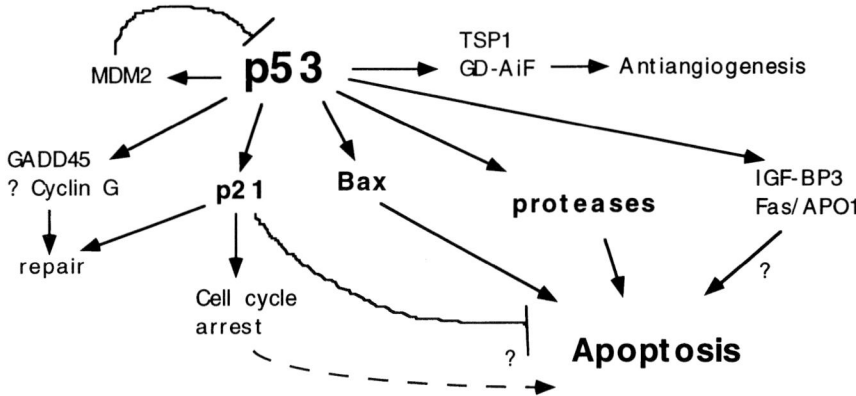

Fig. 3. Transcriptional targets of p53 involved in DNA repair, cell cycle control, growth factor signaling, angiogenesis, and apoptosis

In cancer, p53 function is inactivated through loss of heterozygosity, mutations which can lead to gain of function or act in a dominant negative fashion to inhibit wild-type p53, ubiquitin-mediated degradation through human papillomavirus (HPV) E6-associated protein interaction, MDM2 interaction, and *bax* gene mutation. Normally, p53 protein is stabilized following DNA damage and subsequently cell cycle arrest occurs to permit repair. Defects in the p53 pathway have been associated with increased mutation frequency, "genomic instability" involving chromosomal abberations, tumor development and progression, and chemotherapeutic drug resistance (see Fig. 2).

Fundamental to understanding p53 function is the recognition that p53 regulates not only checkpoint control through p21 induction, but also activates a pathway of apoptotic cell death which inhibits tumor development (see Figs. 2 and 3). It is likely that both of these pathways contribute to the maintenance of genomic stability, and it is likely that defects in each of these pathways contributes to some aspect of the transformed and ultimately the neoplastic phenotype. Much is known about the interactions between p53 and various cellular proteins, as well as the biochemical properties of p53 as a transactivator or repressor of gene expression (LEVINE 1997; VELCULESCU and EL-DEIRY 1996). In addition to the role of p21 in cell cycle arrest and repair, other transcriptional targets of p53 (see Fig. 3) have been linked to growth arrest and repair (GADD45), apoptosis [*bax*, Fas/APO1, insulin-like growth factor (IGF)-BP3], anti-angiogenesis (Thrombospondin 1, GD-AiF), and feedback regulation (MDM2). Recent evidence has suggested that MDM2 not only inhibits transactivation by p53, but also promotes its degradation (HAUPT et al. 1997; KUBBUTAT et al. 1997). MDM2 expression has also been associated with polyploidy (LUNDGREN et al. 1997), as well as p53-independent effects leading to cellular transformation (SIGALAS et al. 1996). p53 appears to activate apoptosis through regulation of proteases that mediate apoptosis (Fig. 3; LEVINE 1997; SABBATINI et al. 1997; WU et al. 1997). The effects of p53 targets, such as p21,

GADD45, *bax* or MDM2 may individually contribute to influence p53's function as guardian of the genome. p53-interacting proteins, such as RAD51 (STURZBECHER et al. 1996) are known to play a role in recombination and repair processes. However, the precise role of RAD51 in maintaining genomic integrity in vivo by p53 remains unclear. It has recently been shown that p53 inactivation leads to high rates of homologous recombination (MEKEEL et al. 1997).

6 p53/p21 and Genomic Instability

Several lines of evidence link p53 to a role in maintaining genomic integrity. Loss or inactivation of p53 leads to gene amplification (LIVINGSTONE et al. 1992; YIN et al. 1992). p53−/− mouse cells (which characteristically escape from senescence) also demonstrate a high degree of aneuploidy (HARVEY et al. 1993). Genomic instability as evidence by aneuploidy, amplifications, and deletions has been observed in tumors derived from p53-deficient/Wnt-1 transgenic mice (DONEHOWER et al. 1995). Human keratinocytes immortalized by HPV and subsequently transformed by *N*-methyl-*N'*-nitro-*N*-nitrosoguanidine (MNNG) exhibited evidence of genetic instability (SHIN et al. 1996). There is some data to suggest the existence of a complex relationship between chromosome instability, aneuploidy, and p53 status, at least in colorectal cancer (LENGAUER et al. 1997). For example, there are chromosomally stable cells that express mutant p53 and chromosomally unstable cells that contain wild-type p53 (LENGAUER et al. 1997). It is possible that other factors, such as MDM2, may contribute to chromosomal instability in wild-type p53-expressing cells, either through wild-type p53 inactivation, degradation, or through p53-independent mechanisms (LUNDGREN et al. 1997; HAUPT et al. 1997; KUBBUTAT et al. 1997).

Wild-type p53 regulates multiple cell cycle checkpoints. Loss of the DNA damage-activated G1 chekpoint has been reported to lead to an increased mutation frequency (HAVRE et al. 1995; BUETTNER et al. 1996; LIU et al. 1996c). It is believed that the increased mutation frequency following UV damage is the consequence of defective nucleotide excision repair, but not transcription-coupled repair (FORD and HANAWALT 1995). Following UV damage, mutant p53-expressing cells inhibit defective repair of cyclobutane pyrimidine dimers in vivo (FORD and HANAWALT 1995). It was proposed that, although the reduced efficiency of repair in cells lacking wild-type p53 may lead to greater genomic instability, resistance to DNA-damaging agents may be enhanced through elimination of apoptosis (FORD and HANAWALT 1995). There is evidence that wild-type p53 plays a role in the elimination of cells with unstable chromosome aberrations (SCHWARTZ and JORDAN 1997) and that the loss of p53 permits survival of clonogenic mutants (GRIFFITHS et al. 1997). However, it was shown that p53-deficiency can lead to increased tumor proliferation and genomic instability, without a significant difference in apoptosis in vivo (HUNDLEY et al. 1997).

Loss of other p53-dependent checkpoints may lead to loss of genomic integrity. Loss of p53 leads to continued DNA synthesis despite mitotic arrest induced by mitotic spindle inhibitors (CROSS et al. 1995). G2/M-phase arrest also occurs following DNA damage of p53-deficient cells which continue to synthesize DNA and become polyploid, a manifestation of S- and M-phase uncoupling (WALDMAN et al. 1996). Loss of p53 leads to centrosome amplification (FUKASAWA et al. 1996). The abnormal amplification of centrosomes impacts mitotic fidelity, leads to unequal segregation of chromosomes, and has been suggested as a key mechanism of genetic instability associated with p53 loss (FUKASAWA et al. 1996). In ataxia-telangiectasia, a disease caused in part by defective p53-dependent checkpoint control (KASTAN et al. 1992), there is evidence of genomic instability and, in fact, lymphoblastic lymphomas derived from ATM-disrupted mice contain karyotypic abnormalities (BARLOW et al. 1996). ATM$-/-$ mice are infertile due to meiotic failure believed to result from abnormal chromosomal synapsis and subsequent chromosome fragmentation (XU et al. 1996). ATM$-/-$ cells exhibit defective cell cycle arrest following radiation which correlates with defective p53 upregulation (XU and BALTIMORE 1996).

While there is plenty of evidence implicating a role for p53 in maintaining genomic integrity, the mechanism by which this is accomplished is not entirely clear. Possibilities include direct effects on repair processes through interactions with various proteins involved in repair of recombination (WANG et al. 1995, 1996; BOGUE et al. 1996; STURZBECHER et al. 1996; SCULLY et al. 1997; SHARAN et al. 1997; HAKEM et al. 1997; MEKEEL et al. 1997), activation of genes involved in cell cycle or cell death pathways (see above) or transcription-independent mechanisms that impact on cell growth (WALKER AND LEVINE 1996). Loss of p53 leading to gene amplification has been associated with loss of p21 expression and cyclin-CDK subunit rearrangement at an early stage in cellular transformation (WHITE et al. 1994; XIONG et al. 1993b, 1996). p21$-/-$ mouse fibroblasts have a deficiency in cell cycle arrest following exposure to an agent that induces CAD gene amplification, PALA (DENG et al. 1995). However, it has not been determined if p21$-/-$ cells have an increased frequency of PALA-resistant colonies due to gene amplification, as compared to p21$+/+$ cells. Thus, whether p21 is required to suppress abnormal gene amplification remains unclear. p53-deficient cells were found to have dramatically increased rates of homologous recombination, but this could not be attributed to p21 (MEKEEL et al. 1997). It was suggested that p53 may suppress recombination in the absence of DNA damage, i.e., as a role to maintain genomic integrity (MEKEEL et al. 1997). p21$-/-$ cells have been reported to have a deficiency in the repair of UV- or cisplatin-induced DNA damage and to have an increase in their spontaneous mutation frequency at the hprt locus (McDONALD et al. 1996). Inhibition of GADD45 expression by antisense strategies has also led to decreased DNA repair and sensitization to UV-irradiation and cisplatin (SMITH et al. 1996). In MNNG-transformed HPV-immortalized human keratinocytes, there was a failure to upregulate either p21 or GADD45 following MNNG exposure, as compared to normal keratinocytes (SHIN et al. 1996). There is evidence that GADD45 and p21 have similar growth suppressive effects (ZHAN et al. 1995) and may coordinately regulate PCNA function (CHEN et al. 1995a). Lastly, a link between MDM2 and polyploidy has recently been uncovered (LUNDGREN et al. 1997). Thus, it is likely that p53's role in maintaining genomic integrity involves multiple downstream targets.

Acknowledgements. Thanks are extended to Paolo Dotto, Steve Elledge, Al Fornace, Renato Iozzo, Phil Leder, David Livingston, Gigi Lozano, Tak Mak, Joe Nevins, Patrick O'Connor, Moshe Oren, Geoff Wahl, and Todd Waldman for providing information relevant to this review. The critical comments of members of my laboratory were appreciated. I sincerely apologize to colleagues if I omitted certain references. I am grateful for support provided by the University of Pennsylvania Comprehensive Cancer Center and the Howard Hughes Medical Institute.

Note added in Proof. Brown et al. (1997) Science 277:831–834 found that the absence of p21 (CIP1/WAF1) is associated with delayed senescence of human fibroblasts. Interestingly, the p21-/- fibroblasts appeared to have an increased rate of telomere shortening. The authors also reported that p21-/- cells did not have an increased rate of CAD gene amplification following PALA selection. Wu et al. (1997) Nature Genetics 17:141–143 report the identification of KILLER/DR5, a novel chemotherapy-inducible p53-regulated cell death receptor gene located on human chromosome 8p21. KILLER/DR5 is a member of the TNF-receptor family and is most closely related to DR4, the receptor for the cytotoxic ligand TRAIL. The authors suggest that the KILLER/DR5 receptor protein may be involved in chemotherapy- and radiation-induced cell death and may contribute to p53-dependent apoptosis through activation of the caspase cascade.

References

Aprelikova O, Xiong Y, Liu ET (1995) Both p16 and p21 families of cyclin-dependent kinase (CDK) inhibitors block the phosphorylation of cyclin-dependent kinases by the CDK-activating kinase. J Biol Chem 270:18195–18197
Bae I, Smith ML, Fornace AJ Jr (1995) Induction of p53-, MDM2- and WAF1/CIP1-like molecules in insect cells by DNA-damaging agents. Exp Cell Res 217:541–545
Balbin M, Hannon GJ, Pendas AM, Ferrando AA, Vizoso F, Fueyo A, Lopez-Otin C (1996) Functional analysis of a p21WAF1/CIP1/SDI1 mutant (arg94-trp) identified in a human breast carcinoma. J Biol Chem 271:15782–15786
Ball KL, Lane DP (1996) Human and plant proliferating-cell nuclear antigen have a highly conserved binding site for the p53-inducible gene product p21WAF1. Eur J Biochem 237:854–861
Barlow C, Hirotsune S, Paylor R, Liyanage M, Eckhaus M, Collins F, Shiloh Y, Crawley JN, Ried T, Tagle D, Wynshaw-Boris A (1996) Atm-Deficient mice: a paradigm of Ataxia Telangiectasia. Cell 86:159–171
Blagosklonny MV, Wu GS, Omura S, El-Deiry WS (1996) Proteasome-dependent regulation of p21WAF1/CIP1 expression. Biochem Biophys Res Commun 227:564–569
Bogue MA, Zhu C, Aguilar-Cordova E, Donehower LA, Roth DB (1996) p53 is required for both radiation-induced differentiation and rescue of V(D)J rearrangement in scid mouse thymocytes. Genes Dev 10:553–565
Bohmer R-M, Scharf E, Assoian RK (1996) Cytoskeletal integrity is required throughout the mitogen stimulation phase of the cell cycle and mediates the anchorage-dependent expression of cyclin D1. Mol Biol Cell 7:101–111
Brugarolas J, Chandrasekaran C, Gordon JI, Beach D, Jacks T, Hannon GJ (1995) Radiation-induced cell cycle arrest compromised by p21 deficiency. Nature 377:552–557
Buettner VL, Hill KA, Nishino H, Schaid DJ, Frisk CS, Sommer SS (1996) Increased mutation frequency and altered spectrum in one of four thymic lymphomas derived from tumor prone p53/Big Blue double transgenic mice. Oncogene 13:2407–2413

Chen IT, Smith ML, O'Connor PM, Fornace AJ Jr (1995a) Direct interaction of Gadd45 with PCNA and evidence for competitive interaction of Gadd45 and p21Waf1/Cip1 with PCNA. Oncogene 11:1931–1937

Chen J, Jackson PK, Kirschner MW, Dutta A (1995b) Separate domains of p21 involved in the inhibition of Cdk kinase and PCNA. Nature 374:386–388

Chen J, Chen S, Saha P, Dutta A (1996) p21Cip1/Waf1 disrupts the recruitment of human Fen1 by proliferating-cell nuclear antigen into the DNA replication complex. Proc Natl Acad Sci USA 93:11597–11602

Chin YE, Kitagawa M, Su W-CS, You Z-H, Iwamoto Y, Fu X-Y (1996) Cell growth arrest and induction of cyclin-dependent kinase inhibitor p21WAF1/CIP1 mediate by STAT1. Science 272:719–722

Cox LS, Midgley CA, Lane DP (1994) Xenopus p53 is biochemically similar to the human tumour suppressor protein p53 and is induced upon DNA damage in somatic cells. Oncogene 9:2951–2959

Cross SM, Sanchez CA, Morgan CA, Schimke MK, Ramel S, Idzerda RL, Raskind WH, Reid BJ (1995) A p53-dependent mouse spindle checkpoint. Science 267:1353–1356

Datto MB, Yu Y, Wang XF (1995) Functional analysis of the transforming growth factor beta responsive elements in the WAF1/Cip1/p21 promoter. J Biol Chem 270:28623–28628

de Nooij JC, Letendre MA, Hariharan IK (1996) A cyclin-dependent kinase inhibitor, Dacapo, is necessary for timely exit from the cell cycle during Drosophila embryogenesis. Cell 87:1237–1247

Deng C, Zhang P, Harper JW, Elledge SJ, Leder P (1995) Mice lacking p21CIP1/WAF1 undergo normal development, but are defective in G1 checkpoint control. Cell 82:675–684

Donehower LA, Godley LA, Aldaz CM, Pyle R, Shi Yp, Pinkel D, Gray J, Bradley A, Medine D, Varmus HE (1995) Deficiency of p53 accelerates mammary tumorigenesis in Wnt-1 transgenic mice and promotes chromosomal instability. Genes Dev 9:882–895

Dulic V, Kaufmann WK, Wilson SJ, Tlsty TD, Lees E, Harper JW, Elledge SJ, Reed SI (1994) p53-dependent inhibition of cyclin-dependent kinase activities in human fibroblasts during radiation-induced G1 arrest. Cell 76:1013–1024

El-Deiry WS (1996) p53, p21WAF1/CIP1 and the control of cell proliferation. In: Thomas NSB (ed) Cell cycle control and apoptosis in malignant disease. Bios Scientific, Oxford, UK, pp 55–75

El-Deiry WS, Tokino T, Velculescu VE, Levy DB, Parsons R, Trent JM, Lin D, Mercer WE, Kinzler KW, Vogelstein B (1993) WAF1, a potential mediator of p53 tumor suppression. Cell 75:817–825

El-Deiry WS, Harper JW, O'Connor PM, Velculescu VE, Canman CE, Jackman J, Pietenpol JA, Burrell M, Hill DE, Wang Y, Wiman KG, Mercer WE, Kinzler KW, Vogelstein B (1994) WAF1/CIP1 is induced in p53-mediated G1 arrest and apoptosis. Cancer Res 54:1169–1174

El-Deiry WS, Tokino T, Waldman T, Velculescu V, Oliner JD, Burell M, Hill DE, Rees JL, Hamilton SR, Kinzler KW, Vogelstein B (1995) Topological control of p21WAF1/CIP1 expression in normal and neoplastic tissues. Cancer Res 55:2910–2919

Elledge SJ (1996) Cell cycle checkpoints: preventing an identity crisis. Science 274:1664–1672

Fan S, Chang JK, Smith ML, Duba D, Fornace AJ Jr, O'Connor PM (1997) Cells lacking CIP1/WAF1 genes exhibit preferential sensitivity to cisplatin and nitrogen mustard. Oncogene 14:2127–2136

Ford JM, Hanawalt PC (1995) Li-Fraumeni syndrome fibroblasts homozygous for p53 mutations are deficient in global DNA repair but exhibit normal transcription-coupled repair and enhanced UV resistance. Proc Natl Acad Sci USA 92:8876–8880

Fukasawa K, Choi T, Kuriyama R, Rulong S, Vande Woude GF (1996) Abnormal centrosome amplification in the absence of p53. Science 271:1744–1747

Givol I, Givol D, Rulong S, Resau J, Tsarfaty I, Hughes SH (1995) Overexpression of human p21waf1/cip1 arrests the growth of chicken embryo firboblasts transformed by individual oncogenes. Oncogene 11:2609–2618

Gorospe M, Wang X, Guyton KZ, Holbrook NJ (1996) Protective role of p21Waf1/Cip1 against prostaglandin A2-mediated apoptosis of human colorectal carcinoma cells. Mol Cell Biol 16:6654–6660

Griffiths SD, Clarke AR, Healy LE, Ross G, Ford AM, Hooper ML, Wyllie AH, Greaves M (1997) Absence of p53 permits propagation of mutant cells following genotoxic damage. Oncogene 14:523–531

Gu Y, Turck CW, Morgan DO (1993) Inhibition of CDK2 activity in vivo by an associated 20K regulatory subunit. Nature 366:707–710

Gulbis JM, Kelman Z, Hurwitz J, O'Donnell M, Kuriyan J (1996) Structure of the C-terminal region of p21WAF1/CIP1 complexed with human PCNA. Cell 87:297–306

Hakem R, Luis de la Pompa J, Elia A, Potter J, Mak TW (1997) Partial rescue of BRCA15-6 early embryonic lethality by p53 or p21 null mutation. Nature Gen 16:298–302

Harper JW, Adami GR, Wei N, Keyomarsi K, Elledge SJ (1993) The p21 cdk-interacting protein Cip1 is a potent inhibitor of G1 cyclin-dependent kinases. Cell 75:805–816

Harper JW, Elledge SJ, Keyomarsi K, Dynlacht B, Tsai L-H, Zhang P, Dobrowolski S, Bai C, Connell-Crowley L, Swindell E, Fox MP, Wei N (1995) Inhibition of cyclin-dependent kinases by p21. Mol Biol Cell 6:387–400

Harvey M, Sands AT, Weiss RS, Hegi ME, Wiseman RW, Pantazis P, Giovanella BC, Tainsky MA, Bradley A, Donehower LA (1993) In vitro growth characteristics of embryo fibroblasts isolated from p53-deficient mice. Oncogene 8:2457–2467

Haupt Y, Maya R, Kazaz A, Oren M (1997) Mdm2 promotes the rapid degradation of p53. Nature 387:296–299

Havre PA, Yuan J, Hedrick L, Cho KR, Glazer PM (1995) p53 inactivation by HPV16 E6 results in increased mutagenesis in human cells. Cancer Res 55:4420–4424

Hundley JE, Koester SK, Troyer DA, Hilsenbeck SG, Subler MA, Windle JJ (1997) Increased tumor proliferation and genomic instability without decreased apoptosis in MMTV-ras mice deficient in p53. Mol Cell Biol 17:723–731

Huppi K, Siwarski D, Dosik J, Michieli P, Chedid M, Reed S, Mock B, Givol D, Mushinski JF (1994) Molecular cloning, sequencing, chromosomal localization and expression of mouse p21 (Waf1). Oncogene 9:3017–3020

Jiang H, Lin J, Su ZZ, Herlyn M, Kerbel RS, Weissman BE, Welch DR, Fisher PB (1995) The melanoma differentiation-associated gene mda-6, which encodes the cyclin-dependent kinase inhibitor p21, is differentially expressed during growth, differentiation and progression in human melanoma cells. Oncogene 10:1855–1864

Kastan MB, Zhan Q, El-Deiry WS, Carrier F, Jacks T, Walsh WV, Plunkett BS, Vogelstein B, Fornace AJ Jr (1992) A mammalian cell cycle checkpoint pathway utilizing p53 and GADD45 is defective in ataxia telangiectasia. Cell 71:587–597

Kubbutat MHG, Jones SN, Vousden KH (1997) Regulation of p53 stability by Mdm2. Nature 387:299–303

LaBaer J, Garrett MD, Stevenson LF, Slingerland JM, Sandhu C, Chou HS, Fattaey A, Harlow E (1997) New functional activities for the p21 family of CDK inhibitors. Genes Dev 11:847–862

Lane ME, Sauer K, Wallace K, Jan YN, Lehner CF, Vaessin H (1996) Dacapo, a cyclin-dependent kinase inhibitor, stops cell proliferation during Drosophila development. Cell 87:1225–1235

Lengauer C, Kinzler KW, Vogelstein B (1997) Genetic instability in colorectal cancers. Nature 386:623–627

Levine AJ (1997) p53, the cellular gatekeeper for growth and division. Cell 88:323–331

Li K-S, Rishi AK, Shao Z-M, Dawson MI, Jong L, Shroot B, Reichert U, Ordonez J, Fontana JA (1996) Posttranscriptional regulation of p21 WAF1/CIP1 expression in human breast carcinoma cells. Cancer Res 56:5055–5062

Li R, Waga S, Hannon GJ, Beach D, Stillman B (1994) Differential effects by the p21 cdk inhibitor on PCNA-dependent DNA replication and repair. Nature 371:534–537

Li W, Fan J, Hochhauser D, Bertino JR (1997) Overexpression of p21waf1 leads to increased inhibition of E2F-1 phosphorylation and sensitivity to anticancer drugs in retinoblastoma-negative human sarcoma cells. Cancer Res 57:2193–2199

Lin J, Reichner C, Wu X, Levine AJ (1996) Analysis of wild-type and mutant p21WAF1 gene activities. Mol Cell Biol 16:1786–1793

Liu M, Lee MH, Cohen M, Bommakanti M, Freedman LP (1996a) Transcriptional activation of the Cdk inhibitor by vitamin D3 leads to the induced differentiation of the myelomonocytic cell line U937. Genes Dev 10:142–153

Liu M, Iavarone A, Freedman LP (1996b) Transcriptional activation of the human p21WAF1/CIP1 gene by retinoic acid receptor. J Biol Chem 271:31723–31728

Liu PK, Kraus E, Wu TA, Strong LC, Tainsky MA (1996c) Analysis of genomic instability in Li-Fraumeni fibroblasts with germline p53 mutations. Oncogene 12:2267–2278

Liu Y, Martindale JL, Gorospe M, Holbrook NJ (1996d) Regulation of p21WAF1/CIP1 expression through mitogen-activated protein kinase signaling pathway. Cancer Res 56:31–35

Livingstone LR, White A, Sprouse J, Livanos E, Jacks T, Tlsty TD (1992) Altered cell cycle arrest and gene amplification potential accompany loss of wild-type p53. Cell 70:923–935

Lundgren K, Montes de Oca Luna R, McNeill YB, Emerick EP, Spencer B, Barfield CR, Lozano G, Rosenberg MP, Finlay CA (1997) Targeted expression of MDM2 uncouples S phase from mitosis and inhibits mammary gland development independent of p53. Genes Dev 11:714–725

Luo Y, Hurwitz J, Massague J (1995) Cell-cycle inhibition by independent CDK and PCNA binding domains in p21Cip1. Nature 375:159–161

Malkowicz SB, Tomaszewski JE, Linnenbach AJ, Cangiano TA, Maruta Y, McGarvey TW (1996) Novel p21WAF1/CIP1 mutations in superficial and invasive transitional cell carcinomas. Oncogene 13:1831–1837

McDonald ER III, Wu GS, Waldman T, El-Deiry WS (1996) Repair defect in p21WAF1/CIP1 −/− human cancer cells. Cancer Res 56:2250–2255

Mekeel KL, Tang W, Kachnic LA, Luo C-M, DeFrank JS, Powell SN (1997) Inactivation of p53 results in high rates of homologous recombination. Oncogene 14:1847–1857

Meng R, Shih H, Prabhu N, George DL, El-Deiry WS (1997) Bypass of abnormal MDM2 inhibition of p53-dependent growth suppression. Clin Cancer Res (in press)

Michieli P, Chedid M, Lin D, Pierce JH, Mercer WE, Givol D (1994) Induction of WAF1/CIP1 by a p53-independent pathway. Cancer Res 54:3391–3395

Michieli P, Li W, Lorenzi MV, Miki T, Zakut R, Givol D, Pierce JH (1996) Inhibition of oncogene-mediated transformation by ectopic expression of p21Waf1 in NIH3T3 cells. Oncogene 12:775–784

Missero C, Di Cunto F, Kiyokawa H, Koff A, Dotto GP (1996) The absence of p21Cip1/WAF1 alters keratinocyte growth and differentiation and promotes ras-tumor progression. Genes Dev 10:3065–3075

Noda A, Ning Y, Venable SF, Pereira-Smith OM, Smith JR (1994) Cloning of senescent cell-derived inhibitors of DNA synthesis using an expression screen. Exp Cell Res 211:90–98

Okuda M, Minehata K, Setoguchi A, Nishigaki K, Watari T, Cevario S, O'Brien SJ, Tsujimoto H, Hasegawa A (1997) Cloning and chromosome mapping of the feline genes p21WAF1 and p27Kip1. Gene (in press)

Parker SB, Eichele G, Zhang P, Rawls A, Sands AT, Bradley A, Olson EN, Harper JW, Elledge SJ (1995) p53-independent expression of p21Cip1 in muscle and other terminally differentiating cells. Science 267:1024–1027

Perkins ND, Felzien LK, Betts JC, Leung K, Beach DH, Nabel GJ (1997) Regulation of NF-κB by cyclin-dependent kinases associated with the p300 coactivator. Science 275:523–527

Polyak K, Waldman T, He T-C, Kinzler KW, Volgelstein B (1996) Genetic determinants of p53-induced apoptosis and growth arrest. Genes Dev 10:1945–1952

Prabhu NS, Blagosklonny MV, Zeng Y-X, Wu GS, Waldman T, El-Deiry WS (1996) Suppression of cancer cell growth by adenovirus expressing p21WAF1/CIP1 deficient in PCNA interaction. Clin Cancer Res 2:1221–1229

Reynisdottir I, Massague J (1997) The subcellular locations of p15Ink4b and p27Kip1 coordinate their inhibitory interactions with cdk4 and cdk2. Genes Dev 11:492–503

Russo AA, Jeffrey PD, Patten AK, Massague J, Pavletich NP (1996) Crystal structure of the p27Kip1 cyclin-dependent-kinase inhibitor bound to the cyclin A-Cdk2 complex. Nature 382:325–331

Sabbatini P, Han J, Chiou S-K, Nicholson DW, White E (1997) Interleukin 1β converting enzyme-like proteases are essential for p53-mediated transcriptionally dependent apoptosis. Cell Growth Differ 8:643–653

Santra M, Mann DM, Mercer WE, Skorski T, Calabretta B, Iozzo RV (1997) Ectopic expression of Decorin protein core causes a generalized growth suppression in neoplastic cells of various histogenetic origin and requires endogenous p21, an inhibitor of cyclin-dependent kinases. J Clin Invest (in press)

Schwartz JL, Jordan R (1997) Selective elimination of human lymphoid cells with unstable chromosome aberrations by p53-dependent apoptosis. Carcinogenesis 18:201–205

Scully R, Chen J, Plug A, Xiao Y, Weaver D, Feunteun J, Ashley T, Livingston DM (1997) Association of BRCA1 with Rad51 in mitotic and meiotic cells. Cell 88:265–275

Serrano M, Lin AW, McCurrach ME, Beach D, Lowe SW (1997) Oncogenic ras provokes premature cell senescence associated with accumulation of p53 and p16INK4A. Cell 88:593–602

Sharan SK, Morimatsu M, Albrecht U, Lim D-S, Regel E, Dinh C, Sands A, Eichele G, Hasty P, Bradley A (1997) Embryonic lethality and radiation hypersensitivity mediated by Rad51 in mice lacking Brca2. Nature 386:804–810

Sheikh MS, Rochefort H, Garcia M (1995) Overexpression of p21WAF1/CIP1 induces growth arrest, giant cell formation and apoptosis in human breast carcinoma cell lines. Oncogene 11:1899–1905

Sheikh MS, Chen YQ, Smith ML, Fornace AJ Jr (1997) Role of p21Waf1/Cip1/Sdi1 in cell death and DNA repair as studied using a tetracycline-inducible system in p53-deficient cells. Oncogene 14:1875–1882

Shim J, Lee H, Park J, Kim H, Choi EJ (1996) A non-enzymatic p21 protein inhibitor of stress-activated protein kinases. Nature 381:804–806

Shin KH, Tannyhill RJ, Liu X, Park NH (1996) Oncogenic transformation of HPV-immortalized human oral keratinocytes is associated with the genetic instability of cells. Oncogene 12:1089–1096

Shiohara M, El-Deiry WS, Wada M, Nakamaki T, Tekeuchi S, Yang R, Chen D-L, Vogelstein B, Koeffler HP (1994) Absence of WAF1 mutations in a variety of human malignancies. Blood 84:3781–3784

Shou W, Dunphy WG (1996) Cell cycle control by Xenopus p28Kix1, a developmentally regulated inhibitor of cyclin-dependent kinases. Mol Biol Cell 7:457–469

Sigalas I, Calvert AH, Anderson JJ, Neal DE, Lunec J (1996) Alternatively spliced mdm2 transcripts with loss of p53 binding domain sequences: transforming ability and frequent detection in human cancer. Nature Med 2:912–917

Smith ML, Kontny HU, Zhan Q, Sreenath A, O'Connor PM, Fornace AJ Jr (1996) Antisense GADD45 expression results in decreased DNA repair and sensitizes cells to u.v.-irradiation or cisplatin. Oncogene 13:2255–2263

Somasundaram K, El-Deiry WS (1997) Inhibition of p53-mediated transactivation and cell cycle arrest by E1A through its p300/CBP-interacting region. Oncogene 14:1047–1057

Somasundaram K, Zhang H, Zeng Y-X, Houvras Y, Peng Y, Zhang H, Wu GS, Licht JD, Weber BL, El-Deiry WS (1997) Arrest of the cell cycle by the tumour suppressor BRCA1 requires the CDK-inhibitor p21Waf1/CipI. Nature 389:187–190

Struzbecher H-W, Donzelmann B, Henning W, Knippschild U, Buchhop S (1996) p53 is linked directly to homologous recombination processes via RAD51/RecA protein interaction. EMBO J 15:1992–2002

Tanaka N, Ishihara M, Lamphier MS, Nozawa H, Matsuyama T, Mak TW, Aizawa S, Tokino T, Oren M, Taniguchi T (1996) Cooperation of the tumor suppressors IRF-1 and p53 in response to DNA damage. Nature 382:816–818

Timchenko NA, Wilde M, Nakanishi M, Smith JR, Darlington GJ (1996) CCAAT/enhancer-binding protein α (C/EBPα) inhibits cell proliferation through the p21 (WAF1/CIP1/SDI-1) protein. Genes Dev 10:804–815

Velculescu VE, El-Deiry WAS (1996) Biological and clinical importance of the p53 tumor suppressor gene. Clin Chem 42:858–868

Waga S, Hannon GJ, Beach D, Stillman (1994) The p21 inhibitor of cyclin-dependent kinases controls DNA replication by interaction with PCNA. Nature 369:574–578

Waldman T, Kinzler KW, Vogelstein B (1995) p21 is necessary for the p53-mediated G1 arrest in human cancer cells. Cancer Res 55:5187–5190

Waldman T, Lengauer C, Kinzler KW, Vogelstein B (1996) Uncoupling of S phase and mitosis induced by anticancer agents in cells lacking p21. Nature 381:713–716

Waldman T et al (1997) Cell cycle arrest vs. cell death in cancer therapy. Nature Med., in press

Walker KK, Levine AJ (1996) Identification of a novel p53 functional domain that is necessary for efficient growth suppression. Proc Natl Acad Sci USA 93:15335–15340

Wang J, Walsh K (1996) Resistance to apoptosis conferred by cdk inhibitors during myocyte differentiation. Science 273:359–361

Wang XW, Yeh H, Schaeffer L, Roy R, Moncollin V, Egly J-M, Wang Z, Friedberg EC, Evans MK, Taffe BG, Bohr VA, Weeda G, Hoeijmakers JHJ, Forrester K, Harris CC (1995) p53 modulation of TFIIH-associated nucleotide excision repair activity. Nature Gen 10:188–195

Wang XW, Vermeulen W, Coursen JD, Gibson M, Lupold SE, Forrester K, Xu G, Elmore L, Yeh H, Hoeijmakers JHJ, Harris CC (1996) The XPB and XPD DNA helicases are components of the p53-mediated apoptosis pathway. Genes Dev 10:1219–1232

White A, Livanos EM, Tlsty TD (1994) Differential disruption of genomic integrity and cell cycle regulation in normal human fibroblasts by the HPV oncoproteins. Genes Dev 8:666–677

Wu GS, Saftig P, Peters C, El-Deiry WS (1997) Potential role for Cathepsin D in p53-dependent tumor suppression and chemosensitivity. Oncogene (in press)

Wu H, Wade M, Krall L, Grisham J, Xiong Y, Van Dyke T (1996) Targeted in vivo expression of the cyclin-dependent kinase inhibitor p21 halts hepatocyte cell-cycle progression, postnatal liver development, and regeneration. Genes Dev 10:245–260

Wyllie FS, Haughton MF, Bond JA, Rowson JM, Jones CJ, Wynford-Thomas D (1996) S phase cell-cycle arrest following DNA damage is independent of the p53/p21WAF1 signaling pathway. Oncogene 12:1077–1082

Xiong Y, Hannon GJ, Zhang H, Casso D, Kobayashi R, Beach D (1993a) p21 is a universal inhibitor of cyclin kinases. Nature 366:701–704

Xiong Y, Zhang H, Beach D (1993b) Subunit rearrangement of the cyclin-dependent kinases is associated with cellular transformation. Genes Dev 7:1572–1583

Xiong Y, Kuppuswamy D, Li Y, Livanos EM, Hixon M, White A, Beach D, Tlsty TD (1996) Alteration of cell cycle kinase complexes in human papilloma E6- and E7-expressing fibroblasts precedes neoplastic transformation. J Virol 70:999–1008

Xu Y, Baltimore D (1996) Dual roles of ATM in the cellular response to radiation and in cell growth control. Genes Dev 10:2401–2410

Xu Y, Ashley T, Brainerd EE, Bronson RT, Meyn MS, Baltimore D (1996) Targeted disruption of ATM leads to growth retardation, chromosomal fragmentation during meiosis, immune defects, and thymic lymphoma. Genes Dev 10:2411–2422

Yin Y, Tainsky MA, Bischoff FZ, Strong LC, Wahl GM (1992) Wild-type p53 restores cell cycle control and inhibits gene amplification in cells with mutant p53 alleles. Cell 70:937–948

Zakut R, Givol D (1995) The tumor suppression function of p21Waf is contained in its N-terminal half ("half-WAF"). Oncogene 11:393–395

Zeng Y-X, El-Deiry WS (1996) Regulation of p21WAF1/CIP1 expression by p53-independent pathways. Oncogene 12:1557–1564

Zeng Y-X, Somasundaram K, El-Deiry WS (1997) AP2 inhibits cancer cell growth and activates p21WAF1/CIP1 expression. Nature Gen 15:78–82

Zhan Q, El-Deiry W, Bae I, Alamo I Jr, Kastan MB, Vogelstein B, Fornace AJ Jr (1995) Similarity of the DNA-damage responsiveness and growth suppressive properties of WAF1/CIP1 and GADD45. Int J Oncol 6:937–946

Zhang H, Hannon GJ, Beach D (1994) p21-containing cyclin kinases exist in both active and inactive states. Genes Dev 8:1750–1758

Zhang W, Grasso L, McLain CD, Gambel AM, Cha Y, Travali S, Deisseroth AB, Mercer WE (1995) p53-independent induction of WAF1/CIP1 in human leukemia cells is correlated with growth arrest accompanying monocyte/macrophage differentiation. Cancer Res 55:668–674

Zhu L, Harlow E, Dynlacht BD (1995) p107 uses a p21 (CIP1)-related domain to bind cyclin/cdk2 and regulate interactions with E2F. Genes Dev 9:1740–1752

Zhu X, Ohtsubo M, Bohmer RM, Roberts JM, Assoian RK (1996) Adhesion-dependent cell cycle progression linked to the expression of cyclin D1, activation of cyclin E-cdk2, and phosphorylation of the retinoblastoma protein. J Cell Biol 133:391–403

Cyclin-Dependent Kinase Inhibitors and Human Cancer

A. KAMB

1 Introduction... 139
2 Somatic Inactivation of CDK Inhibitors............................... 140
3 Germline Mutation of P16.. 141
4 Expression of CDK Inhibitors.. 143
5 Diagnostic Prospects.. 144
6 Therapeutic Implications.. 145
7 Conclusions... 145
References... 146

1 Introduction

As tumor suppressor candidates, cyclin-dependent kinase (CDK) inhibitors have a simple, aesthetic quality. They act directly on the mechanism that drives cells through mitosis and DNA replication. Elimination of CDK inhibitor activity, by mutation for instance, should release CDKs from this form of regulation and remove one of the restraints on cell growth.

As discussed elsewhere in this volume, CDK inhibitors in mammalian cells comprise a family of molecules that, despite similar biochemical properties, appear to have varied functions in the physiology of the cell. All CDK inhibitors discovered to date can be placed in one of two classes based on their peptide sequences: (1) the ankyrin motif or P16 group (P16, P15, P18) and (2) the P21 group (P21, P27, P57). These sequence classifications also correlate with the distinct biochemical behavior of the two groups. The ankyrin CDK inhibitors specifically inhibit CDK4 and CDK6 in vitro. The P21 class is less selective; P21-like molecules inhibit CDK4, CDK6, and CDK2. In both cases, ankyrin inhibitors and p21-like inhibitors binding to CDKs cause inhibition of CDK enzymatic activity in vitro. The role of such inhibition in the cell, however, appears to be different for the various inhibitors.

One manifestation of these differences comes from analysis of the roles of CDK inhibitors in cancer. In the following pages, the current understanding of the

---

Myriad Genetics, Inc., 390 Wakara Way, Salt Lake City, UT 84108, USA

relationship between CDK inhibitors and tumorigenesis will be reviewed. In addition, the prospects of CDK inhibitors as diagnostic and therapeutic tools in cancer will be considered.

2 Somatic Inactivation of CDK Inhibitors

Some of the most significant evidence for the role of specific genes in cancer comes from identification of mutations in tumors. Although this evidence is circumstantial and correlative, in many instances it is compelling. If a candidate tumor suppressor gene can be shown to harbor mutations in a moderate number of tumors, a strong case can be made for the relevance of that gene in tumorigenesis. The premise is that genes which are rendered nonfunctional in neoplasia are inactivated for a reason: they impair tumor growth.

Chromosomal region 9p21, the site of CDK inhibitor genes P15 and P16, is one of the most commonly altered areas in human cancers. For example, loss of heterozygosity (LOH) at 9p21 occurs in over half of bladder tumors and melanomas (TSAI et al. 1990; CAIRNS et al. 1993; DALBAGNI et al. 1993; HEALY et al. 1995; OLAPADE et al. 1990). In addition, homozygous deletions in this region are frequent in both cell lines and primary neoplasia of many kinds (KAMB et al. 1994a; CAIRNS et al. 1995). In the case of cell lines, the rates of deletion are easier to estimate due to the homogeneity of the sample. The homozygous deletion rates average nearly 50% in a wide range of lines. In primary specimens, the reported frequencies of homozygous deletions are variable. Often the measured rates agree between matched primaries and cell lines (CAIRNS et al. 1995). However, in other cases the reported rates are lower in primary samples (KAMB 1995). This may reflect the difficulties of gaining quantitative information about deletions in primary specimens due to the presence of normal tissue and/or the selection in culture of deleted variants.

P16 is certainly one target of the 9p21 alterations (Table 1). P15 may be another target, although it could also be involved merely as a bystander. It lies within a few kilobases of P16. The homozygous deletions center on P16, although P15 is included in the majority. The deletions range in size from a few nucleotides to several megabases.

Table 1. Inactivation of cyclin-dependent kinase (CDK) inhibitor genes in human cancers

Gene	Tumor types screened	Rate	Form of inactivation
p15[a]	Many	High (variable)	Deletion, methylation
p16[a]	Many	High (variable)	Deletion, methylation, point
p18	Leukemia, lung	Low	NF
p21	Several	Low	Point
p27	Many	Low	NF
p57	–	–	–

Point, single base substitution; –, studies have not been reported at the time of writing; NF, none found.
[a] The reported inactivation rates are variable depending on the tumor type and on the particular study.

Although deletions constitute the majority of P16 alterations, point mutations also occur. In pancreatic cancers and melanomas, the rates of point mutation are considerable, i.e. 50% and 10%, respectively (CALDAS et al. 1994; GRUIS et al. 1995). The point mutations are observed almost exclusively in tumors or cell lines that have suffered heterozygous deletion (i.e., LOH) of the other homologue. Interestingly, the point mutations observed in melanoma cell lines have the hallmarks of mutations induced by ultraviolet light, including a high percentage of CC→ TT transitions (LIU et al. 1995). This finding supports the view that the mutations occurred in vivo and are not the result of new mutational events in culture. Extensive screening of the coding sequences and splice junctions of P15 have revealed few somatic mutations (STONE et al. 1995a). This suggests that P15 may be inactivated by deletion in tumors by virtue of its proximity to P16.

In addition to deletion and point mutation, CpG methylation is another mechanism for inactivation of P16 (GONZALEZ-ZULUETA et al. 1995; HERMAN et al. 1995; MERLO et al. 1995). Recent studies have added this new twist to the story of P16, particularly in bladder cancers and leukemias. The percentage of tumors and cell lines with heavily methylated P16 sequences is considerable. In non-small-cell lung carcinoma, for instance, about one quarter of primary tumors have hypermethylated P16 genes. As always with methylation studies, it is difficult to prove whether the increased methylation is the cause or the effect of the transformed status of the cell. However, methylation silences P16 expression, which can be reactivated by treatment with the methylation inhibitor 5-azacytidine. The large number of tumors that contain methylation-silenced P16, along with the other modes of P16 inactivation, lend credence to the idea that methylation is an important mechanism in tumors for eliminating P16 expression. P15 is silenced by methylation at rates that are lower than those for p16, but still appreciable. The best case for a role for P15 in tumorigenesis involves analysis of a set of leukemia tumors and cells lines in which P15, but not P16, is inactivated by methylation (HERMAN et al. 1996).

Remarkably, other CDK inhibitor genes, including P18, P21, P27, and P57, are seldom targets for inactivation in sporadic tumors (Table 1; KAWAMATA et al. 1995; NAKAMIKI et al. 1995; OKAMOTO et al. 1995; VIDAL et al. 1995; WATANABE et al. 1995; C. Cordon-Cardo, personal communication). Efforts to find point mutations or homozygous deletions have yielded few such events. Only in P21 have mutations been identified (albeit at a very low rate). However, it is possible that other modes of inactivation, e.g., methylation, may play a role in a percentage of these genes. But it appears that when it comes to somatic mutation of CDK inhibitor genes in tumors, P16 is exceptional.

3 Germline Mutation of P16

Although P16 was first isolated based on its biochemical properties (SERRANO et al. 1993), its role in cancer was discovered through efforts to identify the familial melanoma gene MLM (KAMB et al. 1994a; NOBORI et al. 1994). Study of mela-

Table 2. Germline sequence variants of p16 gene

Codon	Mutation	Effect	Activity[a]	Arrest[b]
16	delC	Frameshift		
32	T→C	leu→pro		
49	T→C	ile→thr	+/−	
50	A→G	gln→arg		
53	G→C	met→ile		
58	C→T	stop		
60	G→A	ala→thr		
61–2	GC→CG	glu-leu→asp-val		
66	del(191bp)	Frameshift		
71	A→G	asn→ser	+/−	
74	GA→AC	asp→asn	−	−
81		pro→leu	−	
83	C→T	his→tyr	−	
84	G→A	asp→asn	+	+/−
87	G→C	arg→pro	+	+/−
98	A→C	his→pro	+	+/−
100		ala→pro	+	+/−
101	G→T	gly→trp	+/−	+
104–5	del(12bp)	del(4 aa)		
107	G→A	arg→his		
114	C→T	pro→leu	−	−
120	G→A	glu→lys	+	
126	T→A	val→asp	−	
127	G→T	ala→ser	+	+
144	C→T	arg→his		
148	G→A	ala→thr		
IVS2 + 1	G→T	Splice junction		

−, no inhibition; −, does not cause arrest; +/−, intermediate inhibition [a]Inhibitory effect of variants in vitro. [b]Effect of ectopic expression on cultured cells.

noma-prone kindreds from Utah and elsewhere led to localization of MLM to 9p21 (CANNON-ALBRIGHT et al. 1992). When individuals from this set of kindreds and others were screened for mutations in P16, a variety of independent disruptive sequence changes were observed (HUSSUSSIAN et al. 1994; KAMB et al. 1994b). Since these first studies, numerous P16 mutations have been reported in melanoma-prone kindreds around the world (Table 2; SMITH-SORENSON and HOVIG, 1996). The mutations range from microdeletions that cause translational frameshifts in the p16 protein to missense substitutions, several of which have been shown to interfere with CDK inhibition in vitro or in vivo or to destabilize the p16 protein (KOH et al. 1995; LUKAS et al. 1995; RANADE et al. 1995). Surprisingly, some of the best kindreds (i.e., those most likely to segregate 9p21-linked melanoma predisposition) do not appear to have p16-coding sequence mutations (HUSSUSSIAN et al. 1994; KAMB et al. 1994b). Some of these may contain mutations outside the P16-coding sequence. Alternatively, a second 9p21 gene may be involved. Despite its appeal as a candidate for this second gene, P15 is unlikely to account for these missing mutations. Screens of P15 have failed to uncover any germline mutations in melanoma-prone kindreds (STONE et al. 1995a).

What is the overall contribution of P16 mutations to melanoma incidence? This important question can now be answered with reasonable confidence. Several screens of weakly familial melanoma cases and sporadic cases have identified P16 mutations at a very low rate (KAMB et al. 1994b; HOLLAND et al. 1995). These results suggest that P16 germline mutations make a minor contribution to melanoma in general. The vast majority of the disease is either sporadic or caused by predisposition genes with low penetrance. Otherwise, previous genetic studies would have provided evidence for genetic heterogeneity, i.e., the existence of other major loci besides MLM. This has not been the case (CANNON-ALBRIGHT et al. 1992).

Genetic analysis of a set of Dutch melanoma-prone kindreds revealed an astonishing phenomenon. Intermarriage produced two cases of homozygous P16 germline mutations (GRUIS et al. 1995). This seemingly unlikely event, the chance that two heterozygous P16 mutation carriers would marry, provided the opportunity to study a P16 human knockout mutant. The mutation is a 19-bp deletion that destroys protein function. Both homozygotes are viable, but their phenotypes are strikingly different from one another. One survived to the age of 55 with no hint of melanoma, dying apparently of an internal adenocarcinoma. The second suffered three malignant melanomas by the age of 15. Thus the penetrance and expressivity of P16 mutations is highly variable. Penetrance may depend on unidentified genetic factors or on environmental or stochastic influences. Of the potential environmental factors, sun exposure is the most certain (CANNON-ALBRIGHT et al. 1994). Thus P16 is an important predisposing gene for melanoma and possibly other cancers as well, but it is a nonessential gene; it is not required for normal development.

At least one other gene is responsible for a portion of hereditary melanoma. However, it is the exception that proves the rule of P16's importance. The second gene, CDK4, encodes the target of p16's inhibitory activity. Mutations in CDK4 that disrupt the binding of p16 have been found in the germline of a small percentage (two out of 32) of melanoma-prone families (ZUO et al. 1996). Based on this frequency, it appears that CDK4 mutations may account for a fraction of melanoma that is tenfold lower than P16 mutations. The discovery of CDK4 mutations is only the second case of an oncogene's involvement in hereditary cancer. The mutations presumably enable CDK4 to avoid p16-mediated inhibition, resulting in an overactive enzyme that promotes entry into S phase.

So far there is no evidence that other CDK inhibitors contribute to hereditary cancer.

4 Expression of CDK Inhibitors

Besides the loss of P16 expression due to methylation, other types of misregulation occur. The most dramatic case involves P16 expression in Rb-negative tumors. When Rb is inactivated either by mutation or by viral infection, P16 transcript level increases by a factor of 50 (LI et al. 1994; PARRY et al. 1995; STONE et al. 1995b). This suggests that P16 expression is regulated in part by a pathway that involves

Rb. Thus Rb is not only a target of p16, located downstream in the pathway, but is also an upstream negative regulator of P16 expression. This feature of P16 is difficult to reconcile with its transcriptional behavior during the cell cycle. As cells traverse the various phases of the cycle, the level of P16 transcript changes only modestly (STONE et al. 1995b). This also applies to P15 and several other inhibitors (STONE et al. 1995a). If phosphorylation of Rb by CDK is one of the fundamental steps in the G_1/S-phase transition, it is odd that p16 expression is not modulated more. This indicates that p16 may play a minor role in the normal cell cycle oscillator. The relevance of CDK inhibitor expression during normal transitions in the cell cycle remains to be determined.

P27 and P15 have the intriguing property of being regulated by transforming growth factor (TGF)-β (HANNON and BEACH 1994; POLYAK et al. 1994; TOYOSHIMA and HUNTER 1994). This suggests an important role for these genes in G_1-phase arrest mediated by TGF-β. Since escape from TGF-β is a common characteristic of neoplasia, elimination of p27 or p15 activity would seem desirable from the tumor's point of view. Why P27 is not inactivated more frequently in tumors is a mystery.

P21 is one of the transcriptional targets of P53. In many P53-negative tumors, p21 levels are extremely low. This finding led to the proposal that p21-mediated arrest was one of the principal functions of p53 as a tumor suppressor. Despite this tantalizing link, a crucial role for p21 in tumorigenesis has been difficult to prove.

All CDK inhibitors tested to date cause cell cycle arrest when overexpressed in certain tumor cells. In the cases that involve the ankyrin CDK inhibitors, such as p16, the arrest occurs in G_1 phase, as predicted (KOH et al. 1995; LUKAS et al. 1995; SERRANO et al. 1995). Interestingly, p16- and p18-mediated arrest is dependent on functional Rb (GUAN et al. 1994).

5 Diagnostic Prospects

With cancer there are two possible types of gene-based diagnostics: (1) predisposition testing for germline mutations and (2) progression testing for somatic mutations. For technical reasons, germline testing is more straightforward. The mutations are typically heterozygous, but the mutant allele is not obscured by variable amounts of normal DNA. In the case of genes such as BRCA1, the value of germline testing is relatively clear. BRCA1 mutations account for a significant fraction, perhaps 5%, of all breast cancer, the most common cancer that afflicts women.

Testing for germline P16 mutations is more problematic. On the one hand, knowledge of P16 mutational status is useful because it provides an incentive to avoid sunlight, one of the primary modifiers of risk. On the other hand, the low frequency of P16 mutations in the population detracts from the value of a test. The vast majority of individuals would test negative. In addition, the possibility of a large fraction of mutations being of a regulatory nature complicates the test or its interpretation.

Testing for somatic, rather than germline P16 mutations, is more likely to be of value to a large percentage of the population. Somatic P16 mutations are common. Two important issues must be addressed before such tests can be carried out sensibly. First, the test must handle the technical difficulties associated with detecting P16 inactivation by a variety of mechanisms, including homozygous deletion and methylation. Second, the test must be useful either from a prognostic or therapeutic perspective, i.e., the result must provide information about the patient's prognosis or must guide the therapeutic choice. Thus clinical studies that explore the correlation between P16 status and tumor behavior will be welcomed.

It is conceivable that other CDK inhibitors besides p16 may be useful in diagnosis. For example, a protein- or RNA-based detection strategy might provide useful information about the status of growth control pathways in the cell. For instance, if P16 again serves as a paradigm, knowledge of p16 protein levels in tumor tissue might determine whether the cells are negative for P16 or Rb.

6 Therapeutic Implications

One of the wishes harbored by gene therapy enthusiasts is that tumor suppressor genes could be used to complement defects in tumors and restore growth control. The difficulties with this approach are numerous. Perhaps the most significant problems involve delivery of the corrective gene and regulated expression. Since overexpression of p16 would impede growth of normal cells, the expression must be properly controlled.

A more radical approach is to use CDK inhibitors to stop cell division in normal cells selectively and temporarily. The major limitation on success of conventional chemotherapy treatment is its effect on normal cycling cells, such as intestinal epithelia and hematopoietic precursors. These toxic effects severely restrict the doses that can be administered in cancer patients. Some tumor cells often survive and give rise to drug-resistant descendants that ultimately kill the patient. Thus selective protection of normal tissues would likely improve conventional chemotherapy by permitting higher doses of drug. Such approaches have been considered (PARDEE and JAMES 1975; KOHN et al. 1994). In one model system, artificially controlled p16 expression was used to arrest cells in a reversible manner (STONE et al. 1996). In certain cases, the arrested cells were 100-fold more resistant to chemotherapeutic toxicity. After treatment, the arrest could be alleviated by reducing p16 levels, and the cells reentered the cell cycle. This approach, though far from entering practice, offers a different route toward cancer treatment via manipulated CDK inhibitor expression.

7 Conclusions

The role of p16 in human cancer is supported by many lines of evidence, including the following:

- Biochemical inhibition of CDK4/6 activity
- Somatic mutations and deletions in tumors/cell lines
- Methylation-silenced transcription in tumors/cell lines
- Germline mutations in melanoma-prone kindreds
- G_1-phase arrest caused by ectopic expression

Other CDK inhibitors presumably also play roles in the biology of tumors. Methods that do not rely on nature's bounty of mutations to pinpoint genes relevant to cancer must be used to determine these roles. As tumor suppression pathways are dissected in normal cells, these roles should become increasingly clear.

Acknowledgements. I thank Dr. S. Stone and P. Dayananth for comments on the manuscript and F. Bartholomew for terrific assistance.

References

Cairns P, Proctor AJ, Knowles MA (1993) Initiation of bladder cancer may involve deletion of a tumor suppressor gene on chromosome 9. Oncogene 8:1083–1085
Cairns P, Polascik TJ, Eby Y, Tokino K, Califano J, Merlo A, Mao L, Herath J, Jenkins R, Westra W, Rutter JL, Buckler A, Gabrielson E, Tockman M, Cho KR, Hedrick L, Bova GS, Isaacs W, Koch W, Schwab D, Sidransky D (1995) Frequency of homozygous delection at p16/CDKN2 in primary human tumours. Nature Genet 11:210–212
Caldas C, Hahn SA, da Costa LT, Redston MS, Schutte M, Seymour AB, Weinstein CL, Hruban RH, Yeo CJ, Kern SE (1994) Frequent somatic mutations and homozygous deletions of the p16 (MTS1) gene in pancreatic adenocarcinoma. Nature Genet 8:27–32
Cannon-Albright LA, Goldgar G, Meyer LJ, Lewis CM, Anderson DE, Fountain JW, Hegi ME, Wiseman RW, Petty EM, Bale AE, Olopade OI, Diaz MO, Kwiatkowski DJ, Piepkorn MW, Zone JJ, Skolnick MH (1992) Assignment of a melanoma susceptibility locus, MLM, to chromosome 9p13–p22. Science 258:1148–152
Cannon-Albright LA, Meyer LJ, Goldgar DE, Lewis CM, McWhorter WP, Jost M, Harrison D, Anderson DE, Zone JJ, Skolnick MH (1994) Penetrance and expressivity of the chromosome 9p melanoma susceptibility locus (MLM). Cancer Res 54:6041–6044
Dalbagni G, Presti J, Reuter V, Fair WR, Cardon-Cardo C (1993) Genetic alterations in bladder cancer. Lancet 342:469–471
Gonzalez-Zulueta M, Bender CM, Yang AS, Nguyen T, Beart RW, Van Tornout JM, Jones PA (1995) Methylation of the 5' CpG island of the p16/CDKN2 tumor suppressor gene in normal and a transformed human tissues correlates with gene silencing. Cancer Res 55:4531–4535
Gruis NA, Weaver-Feldhaus J, Liu Q, Frye C, Ecles R, Orlow I, Lacombe L, Ponce-Castoneda V, Lianes P, Latres E, Skolnick M, Cardon Cardo C, Kamb A (1995) Genetic evidence in melanoma and bladder cancers that p16 and p53 function in separate pathways of tumor suppression. Am J Pathol 146:1199–1206
Guan K-L, Jenkins CW, Li Y, Nichols MA, Wu X, O'Keefe CL, Matera AG, Xiong Y (1994) Growth suppression by p18, a p16$^{INK4/MTTS1}$ and p14$^{INK4/MTTS1}$ related CDK6 inhibitor, correlates with wild-type pRb function. Genes Dev 8:2939–2952
Hannon GJ, Beach D (1994) p15^{INK4B} is a potential effector of TFG-β-induced cell cycle arrest. Nature 371:257–260
Healy E, Rehman I, Angus B, Rees JL (1995) Loss of heterozygosity in sporadic primary cutaneous melanoma. Genes Chromosomes Cancer 12:152–156
Herman JG, Merlo A, Mao L, Lapidus RG, Issa JJ-P, Davidson NE, Sidransky D, Bayliin SB (1995) Inactivation of the CDKN2/P16/MTS1 gene is frequently associated with aberrant DNA methylation in all common human cancers. Cancer Res 55:4525–4530

Herman JG, Jen J, Merlo A, Baylin SB (1996) Hypermethylation-associated inactivation indicates a tumor suppressor role for p16^{INK4B} Cancer Res 56:722–727

Holland EA, Beaton SC, Becker TM, Grulet OMC, Peters BA, Rizos H, Kefford RF, Mann GJ (1995) Analysis of the p16 gene, CDKN2, in 17 Australian melanoma kindreds. Oncogene 11:2289–2294

Hussussian CJ, Struewing JP, Goldstein A-MN, Higgins PAT, Ally DS, Sheahan MD, Clark WH Jr, Tucker MA, Dracopoli NC (1994) Germline p16 mutations in familial melanoma. Nature Genet 8:15–21

Kamb A (1995) Cell-cycle regulators and cancer. Trends in Genetics 11:136–140

Kamb A, Gruis NA, Weaver-Feldhaus J, Liu Q, Harshman K, Tavtigian SV, Stockert E, Day RS III, Johnson BE, Skolnick MH (1994a) A cell cycle regulator potentially involved in genesis of many tumor types. Science 264:436–440

Kamb A, Shattuck-Eidens D, Eeles R, Liu Q, Gruis NA, Ding W, Hussey C, Tran T, Miki Y, Weaver-Feldhaus J, McClure M, Aitken JF, Anderson DE, Bergman W, Frants R, Goldgar DE, Green A, MacLennan R, Martin NG, Meyer LJ, Youl P, Zone JJ, Skolnick MH, Cannon-Albright LA (1994b) Analysis of CDKN2 (MTS1) as a candidate for the chromosome 9p melanoma susceptibility locus (MLM). Nature Genet 8:22–26

Koh J, Enders GH, Cynlacht BD, Harlow E (1995) Tumor-derived p16 alleles encoding proteins defective in cell-cycle inhibition. Nature 375:506–510

Kohn KW, Jackman J, O'Connor PM (1994) Cell cycle and cancer chemotherapy. J Cell Biochem 54:440–452

Kawamata N, Morosetti R, Miller CW, Park D, Spirin KS, Nakamaki T, Takeuchi S, Hatta Y, Simpson J, Wilczynski S, Lee YY, Bartram CR, Koeffler HP (1995) Molecular analysis of the cyclin-dependent kinase inhibitor gene p27/Kip1 in human malignancies. Cancer Res 55:2266–2269

Li Y, Nichols MA, Shay JW, Xiong Y (1994) Transcriptional repression of the D-type cyclin-dependent kinase inhibitor p16 by the retinoblastoma susceptibility gene product pRb. Cancer Res 54:6078–6082

Liu Q, Neuhausen S, McClure M, Frye C, Weaver-Feldhaus J, Gruis NA, Eddington K, Allalunis-Turner MJ, Skolnick MH, Fujimura FK, Kamb A (1995) CDKN2(MTS1) tumor suppressor gene mutations in human tumor cell lines. Oncogene 10:1061–1067

Lukas J, Parry D, Aagaard L, Mann DJ, Bartkova J, Strauss M, Peters G, Bartek J (1995) Retinoblastoma-protein-dependent cell-cycle inhibition by the tumor suppressor p16. Nature 375:503–506

Merlo A, Herman JG, Mao L, Lee DJ, Gabrielson E, Burger PC, Baylin SB, Sidransky D (1995) 5' CpG island methylation is associated with transcriptional silencing of the tumour suppressor p16/CDKN2/MTS1 in human cancers. Nature Med 1:686–691

Nakamaki T, Kawamata N, Schwaller J, Tobler A, Fey M, Pakkala S, Lee YY, Kim BK, Fukuchi K, Tsuruoka N, Kahan J, Miller CW, Koeffler HP (1995) Structural integrity of the cyclin-dependent kinase inhibitor genes, p15, p16 and p18 in myeloid leukaemias Br J Haematol 91:139–149

Nobori T, Miura K, Wu DJ, Lois A, Takabayashi K, Carson DA (1994) Deletions of the cyclin-dependent kinase-4 inhibitor gene in multiple human cancers. Nature 368:753–756

Okamoto A, Hussain SP, Hagiwara K, Spillare EA, Rusin MR, Demetrick DJ, Serrano M, Hannon GJ, Shiseki M, Zariwala M, Xiong Y, Beach DH, Yokota J, Harris CC (1995) Mutations in the p16^{INK4}/MTS1/CDKN2, p15^{INK4B}/MTS2, and p18 genes in primary and metastatic lung cancer. Cancer Res 55:1448–1451

Olopade OI, Jenkins R, Linnenbach AJ et al (1990) Molecular analysis of chromosome 9p deletion in human solid tumors. Proc Am Assoc Cancer Res 21:318

Pardee AB, James LJ (1975) Selective killing of transformed baby hamster kidney (BHK) cells. Proc Natl Acad Sci USA 72:4994–4998

Parry D, Bates S, Mann DJ, Peters G (1995) Lack of cyclin D-Cdk complexes in Rb-negative cells correlates with high levels of p16$^{INK4/MTS1}$ tumour suppressor gene product. EMBO J 14:503–511

Polyak K, Lee M-H, Erdjument-Bromage H, Koff A, Roberts JM, Tempst P, Massague J (1994) Cloning of p27^{Kip1}, a cyclin-dependent kinase inhibitor and a potential mediator of extracellular antimitogenic signals. Cell 78:59–66

Ranade K, Hussussian CJ, Sikorski RS, Varmus HE, Goldstein AM, Tucker MA, Serrano M, Hannon GJ, Beach D, Dracopoli NC (1995) Mutations associated with familial melanoma impair p16^{INK4} function. Nature Genet 10:114–116

Serrano M, Hannon GJ, Beach D (1993) A new regulatory motif in cell-cycle control causing specific inhibition of cyclin D/CDK4. Nature 366:704–707

Serrano M, Gomez-Lahoz E, DePinho RA, Beach D, Bar-Sagi D (1995) Inhibition of Ras-induced proliferation and cellular transformation by p16^{INK4}. Science 267:249–252

Smith-Sorenson B, Hovig E (1996) CDKN2A (p16^{INK4A}) somatic and germline mutations. Hum mutat 7:294–303

Stone S, Dayananth P, Jiang P, Weaver-Feldhaus JM, Tavtigian SV, Cannon-Albright L, Kamb A (1995a) Genomic structure, expression and mutational analysis of the p15 (MTS2) gene. Oncogene 11:987–991

Stone S, Jiang P, Dayananth P, Tavtigian SV, Katcher H, Parry D, Peters G, Kamb A (1995b) Complex structure and regulation of the p16 (MTS1) locus. Cancer Res 55:2988–2994

Stone S, Dayananth P, Kamb A (1996) Reversible p16-mediated cell cycle arrest as protection from chemotherapy. Cancer Res 56:3199–3202

Toyoshima H, Hunter T (1994) p27, a novel inhibitor of G1 cyclin-Cdk protein kinase activity, is related to p21. Cell 78:67–74

Tsai YC, Nichols PW, Hiti AL, Williams Z, Skinner DG, Jones PA (1990) Allelic losses of chromosomes 9,11, and 17 in human bladder cancer. Cancer Res 50:44–47

Vidal MJ, Loganzo F, Oliveira AR, Hayward NK, Albino AP (1995) Mutations and defective expression of the WAF1 p21 tumor suppressor gene in malignant melanomas. Melanoma Res 5:243–250

Watanabe H, Fukuchi K, Takagi Y, Tomoyasu N, Gomi K (1995) Molecular analysis of the Cip1/Wafl (p21) gene in diverse types of human tumors. Biochim Biophys Acta 1263:275–280

Zuo L, Weger J, Yang Q, Goldstein AM, Tucker MA, Walker GJ, Hayward N, Dracopoli NC (1996) Germline mutations in the p16^{INK4a} binding domain of CDK4 in familial melanoma. Nature Genet 12:97–99

Small-Molecule Inhibitors of Cyclin-Dependent Kinases: Molecular Tools and Potential Therapeutics

D.H. WALKER

1	Chemical Inhibitors of Cyclin-Dependent Kinases	149
1.1	Purine Analogues	151
1.2	Flavonoids	153
1.3	Butyrolactone-1	154
1.4	Staurosporine Analogues	154
1.5	Other Inhibitors	155
2	Other Classes of CDK-Inhibitory Compounds	156
2.1	Peptide Inhibitors of CDKs	156
2.2	Inhibitors of cdc25	157
2.3	Antisense Inhibitors of CDKs	157
3	CDK Inhibitors as Molecular Tools	158
4	CDK Inhibitors as Therapies	161
	References	162

1 Chemical Inhibitors of Cyclin-Dependent Kinases

The cell has evolved a number of distinct ways to regulate cyclin-dependent kinases (CDKs). These include both activating steps (phosphorylation at T161, binding of a cyclin) and inhibitory ones (phosphorylation of T14, Y15, binding of a protein inhibitor, and ubiquitin-mediated proteolysis of the cyclin subunit). All represent potential points of intervention for the design of small-molecule inhibitors. Thus, in theory, it will be possible to identify inhibitors of CDK activity which act though a number of different mechanisms, i.e., competition for substrates (ATP and peptide/protein), competition for cyclin binding, mimicking of CDK-inhibitory proteins, or stimulation of specific cyclin destruction. In practice, however, things are not so simple. In order for a small molecule to penetrate the cell membrane, it typically has to be of small size (e.g., M_r < 500). With such size constraints, the highest likelihood is that such a small molecule will bind most tightly in small, defined pockets in the target protein, such as the ATP-binding pocket. The potential for disruption of a large protein–protein interface, such as a cyclin–CDK-binding surface, is much lower. It is therefore predictable that the majority of small molecular weight CDK-inhibitory compounds identified to date are ATP competitors. These fall into

---

Department of Cancer Biology, Glaxo Wellcome, 5 Moore Drive, RTP NC 27709, USA

several major classes, including the purine analogues, the flavonoids, staurosporine analogues, and butyrolactone-1. Despite the fact that these inhibitors currently are either nonselective or low in potency, they all present potential starting points for the identification of highly potent, specific CDK inhibitors. The structures of these inhibitors are shown in Fig. 1 and they are described in more detail below.

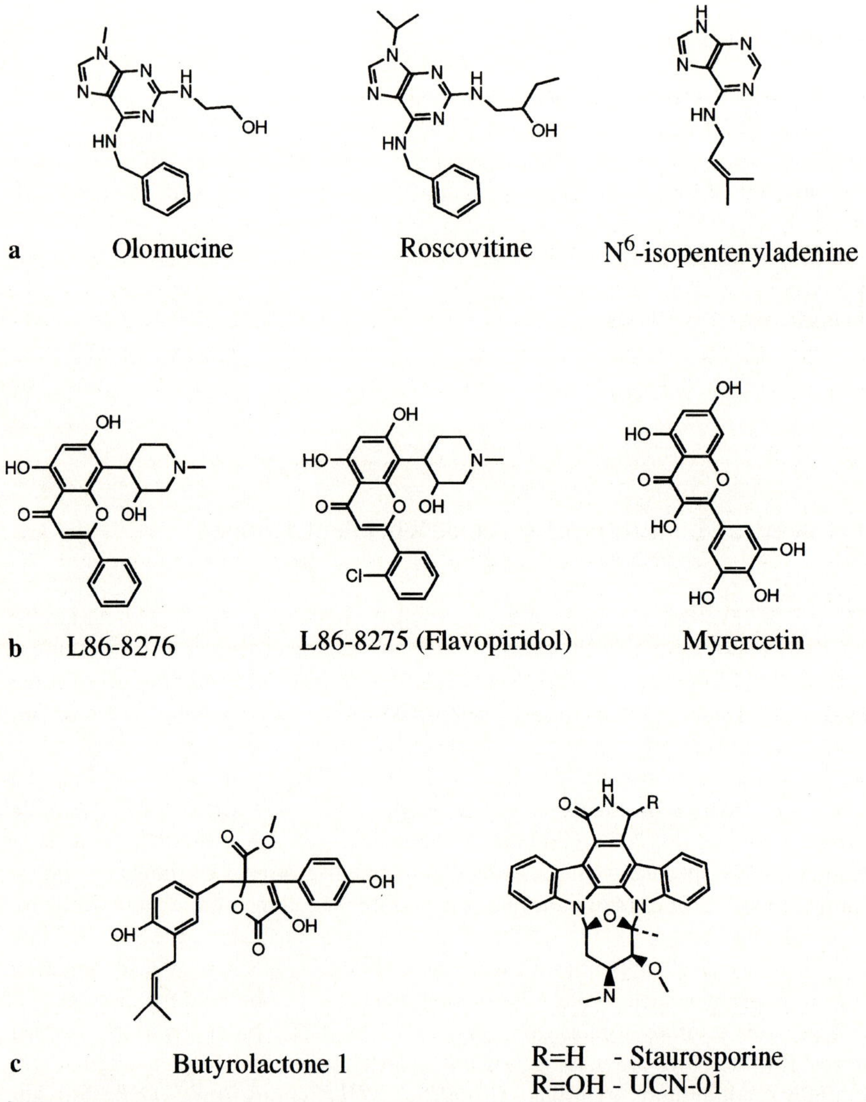

Fig. 1a–c. Small-molecule inhibitors of cyclin-dependent kinases (CDK). The structures of the major CDK inhibitors described in the text are shown. **a** Purines. **b** Flavonoids. **c** Other compound classes. (For the CDK-inhibitory activity of these compounds, see Table 1)

It should be noted that the concentration of ATP used for in vitro studies needs to be taken into account when comparing the in vitro and cellular activities of CDK inhibitors, since differences in [ATP] used will affect the reported IC_{50} for enzyme inhibition. Thus two equipotent inhibitors may appear to possess greatly altered activity simply by assaying them at different [ATP]. The only way to accurately compare these compounds is through the determination of a K_i value. Unfortunately, this has not been performed for many of the compounds described here.

1.1 Purine Analogues

It has been known for several years that purine analogues, such as N6-(D2-isopenentyl) adenine (isopentenyladenine) and 6-dimethylaminopurine (DMAP) have CDK-inhibitory activity (RIALET and MEIJER 1991). These compounds suffer from the problem of low potency and poor selectivity between CDKs and different kinases. In an effort to identify purine analogues with greater potency and selectivity for CDKs, Meijer and coworkers screened more than 80 related purine analogues (VESELY et al. 1994). This search identified olomoucine, 2-(2-hydroxyethylamino)-6-benxylamino-9-methylpurine (Fig. 1, Table 1). This compound showed selectivity for cdc2 (CDK1), CDK2, and CDK5 (IC_{50}, 3–7 μM), with some selectivity for members of the Erk family (IC_{50}, approximately 30 μM) (VESELY et al. 1994). Olo-

Table 1. Activity of small-molecule inhibitors of cyclin-dependent kinases (CDK)

IC_{50} values of inhibitors (μM)

Kinase	Olomoucine	Roscovitine	6-Isopentenyl adenine	Flavopiridol	Butyrolactone-1	Staurosporine	UCN-01
cdc2	7	0.65	40	0.3^a	0.6	0.003–0.009	0.031
CDK2	7	0.7	–	0.4	1.5	0.007	0.03
CDK4	>100	>1000	–	0.4 (0.065^b)	–	<10	0.032
MAPK	30	30	–	–	94	–	0.91
PKA	>1000	>2000	50	145	260	0.007	–
PKC	>100	>1000	–	–	160	0.007	0.007
PKG	>1000	>2000	–	6	–	0.009	–
Src TK	–	–	–	–	–	0.006	–
EGF-R/K	–	440	–	25	>590	0.63	–

Note that the IC_{50} values are affected by the [ATP] used in each assay, making direct comparison of inhibitor activity difficult. The $K_{i\ app}$ can be determined according to the equation: $K_{i\ app} = IC_{50}/(1 + (s/K_m))$. If the [ATP] used is well below the K_m for the enzyme (approximately 40 μM for CDK2), then the IC_{50} is approximately equal to the $K_{i\ app}$. For cdc2, the $K_{i\ app}$ for olomoucine and flavopiridol will be 5 μM and 0.028 μM, respectively.
^{a}[ATP] = 375 μM.
^{b}Reported K_i value.

moucine has little activity against a large number of other kinases and ATP-utilizing enzymes, including CDK4 and CDK6. The extensive list of purines screened, along with the generally good accessibility of purines to chemical modification, provides a good starting point for further improvement of these molecules. For instance, only C2-, N6-, N9-substituted purines demonstrated a strong inhibitory effect towards CDKs. As expected, olomoucine was shown to be competitive with ATP and noncompetitive with the peptide substrate histone H_1. Olomoucine also shows activity in inhibiting growth and cell division in a number of living systems (GLAB et al. 1994; VESELY et al. 1994; ABRAHAM et al. 1995), consistent with the inhibition of CDK activity. These include murine, *Xenopus*, and echinoderm oocytes, unicellular algae, plant cells, and normal and tumor cell lines in tissue culture. In many of these systems, olomoucine shows reversible effects, with a reentry into the cell cycle occurring after removal of the compound. Interestingly, olomoucine does not show any effects in *Drosophila* embryos and yeast and is only weakly active in nematode embryos. The potential reasons for inactivity of olomoucine in these systems are many, but a high concentration of ATP (which will compete with the compound) and its inability to penetrate the cell wall or membrane are likely possibilities. The average IC_{50} for cell growth for olomoucine in the NCI human tumor cell line panel was 60.3 μM (ten fold greater than the IC_{50} for the purified enzyme). This apparent disparity in the activity of olomoucine between the purified enzyme and whole cells may be explained by the fact that the concentration of ATP in cells is very high (>0.5 mM) relative to that used in the kinase assays for testing olomoucine (15 μM). Since olomoucine competes for ATP, higher concentrations of the drug will be required to inhibit the enzyme in the presence of a high substrate concentration.

The crystal structures of isopentenyladenine and olomoucine complexed with CDK2 have been solved (SCHULZE-GAHMEN et al. 1995). These structures confirm that both isopentenyladenine and olomoucine do indeed bind in the ATP pocket, but surprisingly adopt orientations which are different both from the purine ring of ATP itself as well as from each other. The crystal structure of olomoucine has provided two other useful pieces of information. First, the benzylamino group of olomoucine protrudes out of the ATP-binding pocket into the solvent interface. Modeling olomoucine into the ATP pocket of the cyclin-dependent protein kinase (PKA) leads to the overlap of a phenylalanine side chain in PKA with the benzylamino group of olomoucine. This steric hindrance would obviously decrease the affinity of olomoucine for the ATP pocket of PKA and provides one model by which selectivity for CDKs may be generated. It should be noted that this does not provide a model in which CDK4 and CDK6 are insensitive to olomoucine. It is likely that the answer to this puzzle will only be determined by the solution of the crystal structure of CDK4 or CDK6. Second, the crystal structure of olomoucine bound to CDK2 has provided a starting point for further chemical modification of olomoucine. This effort has yielded a further analogue of olomoucine, roscovitine (MEIJER 1996; L. Meijer, personal communication; see Fig. 1). Roscovitine shows an increased potency for cdc2 and CDK2 of at least tenfold, with an IC_{50} for cdc2 of 650 nM (Table 1). Like olomoucine, roscovitine is also inactive against CDK4. Roscovitine has a single chiral center (see Fig. 1). The (*R*)-stereoisomer is about twofold more potent than the (*S*)-

stereoisomer. The improvement in potency for CDKs is not reflected in the activity against other kinases, further improving the selectivity of this inhibitor for CDKs. The improved activity of roscovitine against CDKs is also reflected in improved activity in cell-based systems, e.g., the average IC$_{50}$ in the NCI cell line panel is 16 μM (L. Meijer, personal communication). The use of the crystal structure to identify potent analogues of olomoucine serves to highlight the usefulness of the crystal structure of a protein and molecular modeling in the design of inhibitory compounds. It is likely that further improvements over roscovitine will be made in the near future.

1.2 Flavonoids

The flavonoids have long been known to have kinase-inhibitory properties and many show cell cycle modulation and growth-arresting properties (OGAWARA et al. 1989). Flavopiridol (L86-8275) [(-)cis-5,7-dihydroxy-2-(2-chlorophenyl)-8[4-(3-hydroxy-1-methyl-0-piperidinyl]-4H-benzopyran-4-one] (Fig. 1) and the closely related compound L86-8276 are the most potent flavonoid CDK inhibitors identified to date (LOSIEWICZ et al. 1994). Flavopiridol potently inhibits sea star cdc2 (with a K_{iATP} of 41 nM; LOSIEWICZ et al. 1994) and recombinant human CDK4 (K_{iATP}, 65 nM; CARLSON et al. 1996) (Table 1). The IC$_{50}$ for cdc2 immunoprecipitated from MDA MB468 cells was 400 nM at an ATP concentration of 375 nM (LOSIEWICZ et al. 1994). These data indicate that flavopiridol, unlike olomoucine and butyrolactone (see below), is unable to distinguish between the different CDK family members. Flavopiridol also shows some kinase selectivity, being considerably less potent against a number of other kinases, including PKA, PKG, and the epidermal growth factor (EGF) receptor/kinase. Like olomoucine, flavopiridol is competitive with respect to ATP and non-competitive with respect to the peptide substrate (LOSIEWICZ et al. 1994).

The reported IC$_{50}$ for cell growth – (based on the MTT assay) – for flavopiridol ranges from 25 to 160 nM for a selection of 12 breast and lung carcinoma cell lines (KAUR et al. 1992). Flavopiridol was initially shown to be cytostatic on a variety of tumour cell lines (KAUR et al. 1992; WORLAND et al. 1993; LOSIEWICZ et al. 1994). Recent reports have also shown that flavopiridol can be cytotoxic on certain cell lines (BIBLE and KAUFMANN 1996; PARKER et al. 1996). It has been argued that, based on the selectivity and potency of flavopiridol, the effects of this compound on the growth of cells may be accounted for by the CDK-inhibitory properties of the compound (LOSIEWICZ et al. 1994). In one instance, however (BIBLE and KAUFMANN 1996), flavopiridol has been shown to be cytotoxic against quiescent A549 non-small-cell lung cancer cells. This observation raises some questions regarding the absolute specificity of this compound. Further studies will be needed to determine the degree to which the cytostatic and cytotoxic effects of flavopiridol are related to CDK inhibition or to other properties of the compound.

The crystal structure of CDK2 in complex with L86-8276 has also been solved (AZEVEDO et al. 1996). The benzopyran ring of L86-8276 occupies a similar region as the purine ring of ATP, and the piperidinyl ring of L868276 partially occupies the space in which the α-phosphate of ATP resides. Similarly to olomoucine, the phenyl group of L868276 projects out of the ATP-binding pocket towards the solvent interface, suggesting in part that this may contribute to the specificity of this molecule.

We (B. Lovejoy and D.H. Walker, unpublished data) have solved the crystal structure of the flavonoid myrecetin complexed with CDK2. Myrecetin (see Fig. 1) is a weak inhibitor of CDK2 relative to flavopiridol (IC$_{50}$, 10 μM; [ATP], 40 μM; (D.H. Walker and B. Lovejoy, unpublished results), but represents a further flavonoid structure. Initial crystallographic studies suggest that it is possible for myrecetin to bind CDK2 in a reverse orientation as compared to flavopiridol. These results mirror the observation that purine analogues bind CDK2 differently from ATP and serve to highlight the potential importance of using crystallography to aid in compound design. Without this valuable information, it may be difficult to understand the structure–activity relationships of different inhibitors even from the same class of compounds. The crystal structure of CDK2 in complex with cyclin A has also been solved (JEFFREY et al. 1995). A number of dramatic changes in the CDK2 molecule take place upon binding cyclin A, and these may impact on the binding of small molecules in the ATP pocket. Although in this structure the binding pocket for the adenine portion of ATP is little altered, the changes in CDK2 in the phosphate-binding region may allow the formation of new contacts with the molecule as compared to the free CDK structure. The use of the CDK2–cyclin A structure will allow for a more accurate determination of the critical interactions outside of the nucleotide-binding pocket and will likely significantly impact efforts to improve potency of CDK inhibitors.

1.3 Butyrolactone-1

Butyrolactone-1 was isolated from *Aspergillus* strain F-25799 as a CDK inhibitor in a screen of culture mediums from microorganisms to find inhibitors of murine cdc2 (KITIGAWA et al. 1993). Butyrolactone-1 is competitive with ATP, inhibits cdc2 with an IC$_{50}$ of 680 nM, and shows good specificity against a variety of other kinases, including CDK4 (KITIGAWA et al. 1993; MEIJER 1995; see Table 1). Evidence has also been provided to show that butyrolactone has activity in cell-based assays. It arrests *Xenopus* egg extracts in G$_2$ phase with an IC$_{50}$ of less than 3 μM, consistent with inhibition of cdc2 (SOMEYA et al. 1994), and arrests human W138 fetal lung fibroblasts in both G$_1$ and G$_2$ phase at 48 μM. There are no reports of structural modifications of butyrolactone, and this molecule has not been crystallized with CDK2. Nonetheless, it provides a very attractive template for the design of further analogues, in particular because it represents a class of molecules which have not been previously identified in general kinase screening approaches.

1.4 Staurosporine Analogues

Staurosporine is a microbial alkaloid which was originally isolated from *Streptomyces* sp. cultures (OMURA et al. 1997). Staurosporine is recognized as a potent inhibitor of a wide variety of kinases, including both serine/threonine and tyrosine kinases, and inhibits cdc2 and CDK2 with an IC_{50} in the low nanomolar range (RIALET and MEIJER 1991; GADBOIS et al. 1992; see Table 1). This is in the same range as inhibition of PKC (2.7 nM; RIALET and MEIJER 1991; GADBOIS et al. 1992). Staurosporine has been reported to be poorly active against CDK4–cyclin complexes (MEIJER 1995). Recently, an analogue of staurosporine, 7-hydroxyl-staurosporine (UCN-01) (TAKAHASHI et al. 1989), has been shown to have potent activity against CDKs (KAWAKAMI et al. 1996). Although UCN-01 is somewhat less potent than staurosporine (IC_{50} cdc2 of 30 nM as compared to 4 nM), it shows a marginally improved selectivity and also inhibits CDK4 and CDK6 (KAWAKAMI et al. 1996). The lack of specificity of staurosporine and UCN-01 limits their usefulness as tools for understanding CDK regulation. Thus, although there are many reports of staurosporine or UCN-01 causing cell cycle changes in a variety of target cells (KAWAKAMI et al. 1996; AKINAGA et al. 1994; SCHNIER et al. 1994; SEYNAEVE et al. 1993), it is not possible to determine whether this effect is a result of inhibition of CDKs or other kinases which either directly or indirectly impact cell cycle progression. This problem is further highlighted by reports claiming both inhibition of CDK activity (KAWAKAMI et al. 1996) and activation of CDK activity (WANG et al. 1995) by UCN-01 at similar concentrations in different cell types. It is possible that more specific analogues of staurosporine may be derived, particularly if the crystal structure of CDK2 complexed with staurosporine is solved. The compound is rather large and complex, however, making chemical modification more difficult as compared to other structural classes described here.

1.5 Other Inhibitors

There are a number of other reported CDK inhibitors. Typically, these inhibitors are non-selective and represent molecules in which no structure–activity relationships have been generated. As such, they are of little value themselves as tools or agents for the specific inhibition of CDK activity, but do represent possible, although difficult, templates for further chemical modification for the identification of more selective and potent analogues. These molecules include suramin (BOJANOWSKI et al. 1994) and 9-hydroxy-elipticine (OHASHI et al. 1995). Suramin is currently used as an antihelminthic and antiprotozolal compound and as an antitumor agent (LARSEN 1993). Suramin inhibits cdc2 with an IC_{50} of 4 µM, well above the IC_{50} for other actions of this compound (BOJANOWSKI et al. 1994). Given the complexity of suramin's structure, it is unlikely that this compound will be seriously pursued as a CDK inhibitor. Ellipticines are widely recognized as anti-cancer agents acting as DNA intercalators and inhibitors of topoisomerase II. The potency of these compounds in cell-based assays is well below the reported IC_{50} for CDK inhibition (OHASHI et al. 1995). Since these compounds show DNA-intercalating activity, it is highly unlikely they will provide an adequate template for CDK inhibitors.

2 Other Classes of CDK-Inhibitory Compounds

All of the CDK inhibitors described above represent small organic molecules which are generally competitive with respect to ATP. Identification of peptide CKD inhibitors, inhibition of regulators of CDK activity, and inhibition of transcription of the CDK protein by antisense technologies all represent alternative approaches to inhibition of CDK activity which deserve a mention here. By moving the focus away from the ATP-binding pocket of CDK, these approaches are likely to provide inhibitors with an improved degree of selectivity relative to the inhibitors described above. Nonetheless, they still represent nascent technologies and carry their own individual problems which need to be overcome.

2.1 Peptide Inhibitors of CDKs

The design of peptide CDK inhibitors has been approached by two methods: (1) identifying peptide aptamers which bind and inhibit CDK activity and (2) designing peptide inhibitors based on the existing CDK-inhibitory proteins.

Brent and colleagues (COLAS et al. 1996) used the two-hybrid system to express a combinatorial library of constrained 20-mer peptides displayed in the context of the active-site loop of *Escherichia coli* thioredoxin. This approach has yielded a range of peptides which have high affinity for different epitopes on the CDK2 surface and which inhibit CDK2–cyclin E kinase activity in vitro. For this technology to be of wider interest, it will be necessary to demonstrate that the peptides have CDK-binding activity alone and not only in the constraint of thioredoxin.

An alternative approach used to identifying CDK-inhibitory peptides is to map active regions of known CDK-inhibitory proteins, reducing the inhibitory domain to a small fragment. The protein inhibitors of CDKs fall into two distinct classes: universal inhibitors (such as $p21^{WAF1/CIP1}$ and $p27^{KIP1}$) and the CDK4/6-specific inhibitors, such as $p16^{INK4/MTS1}$. The universal inhibitors make contacts with both the cyclin and the CDK subunits of the majority of cyclin– CDK complexes (RUSSO et al. 1996). $p16^{INK4/MTS1}$, on the other hand, binds the free CDK4/6 subunit. Although domains of $p21^{WAF1/CIP1}$ which mimic their CDK-inhibitory action have been identified (CHEN et al. 1995), these have not yet led to the identification of small inhibitory peptides. More success has been reached in mapping $p16^{INK4/MTS1}$. Lane and coworkers (FAHRAEUS et al. 1996) mapped the CDK-binding region of $p16^{INK4/MTS1}$ to a 20-amino acid peptide which shows CDK4-inhibitory activity. This peptide was coupled to Penetratin, a 16-amino acid region of the Antennapedia homeodomain, which rapidly translocates through biological membranes in an energy-independent manner. The p16 peptide–Penetratin fusion induced a G_1 phase cell cycle arrest and inhibited the phosphorylation of Rb in vivo in HaCaT cells. The activity of the peptide was limited to times prior to the phosphorylation of Rb at the restriction point, suggesting that the inhibitory action of the peptide was associated with inhibition of CDK4/CDK6. The high degree of selectivity of this peptide for CDK4 and CDK6 suggests it as an interesting starting point for the design of cdk4/cdk6-specific inhibitors.

2.2 Inhibitors of cdc25

A major point of regulation of CDKs, particularly in G_2/M phase, is via phosphorylation/dephosphorylation of T14 and Y15 in the CDK by the Wee1 kinase/cdc25 phosphatase pathway (for a review, see HUNTER 1995). Overexpression of cdc25 may be expected to prevent the Wee1-mediated inhibition of CDK and to result in the inappropriate activation of CDKs in the cell. Overexpression of cdc25 has been reported to be found in some primary tumors, such as breast and lung (GALAKTIONOV et al. 1995). This confirms the hypothesis that inappropriate CDK activity may contribute to the neoplastic phenotype. These results further suggest that targeting cdc25 may be a good way of inhibiting CDK activity in the context of some tumors by swinging the balance towards inhibitory phosphorylation, thereby preventing the activation of CDKs.

To date, few inhibitors of cdc25 have been reported. The benzoquinoid antitumor compounds dnacinA1 and dnacin B1 have been shown to be noncompetitive inhibitors of cdc25B (HORIGUCHI et al. 1994). However, the activity of these compounds is very weak (IC_{50}, 141 and 64.4 μM, respectively). Dysidiolide, a sesterterpine γ-hydroxybutenolide, which was isolated from the marine sponge *Dysidea etheria* (GUNASEKERA et al. 1996), shows somewhat better potency. Dysideolide inhibits cdc25 with an IC_{50} of 9.4 μM and is inactive against calcineurin, CDK45, and leukocyte common antigen-related phosphatase (LAR) at 12.5 μM, indicating the potential for specificity of this molecule. Since phosphatase inhibitors do not have to contend with high levels of ATP within the cell, as the ATP-competitive kinase inhibitors do, they are more likely to show cellular effects at concentrations near the reported IC_{50} values. Dysidiolide causes growth inhibition of A549 human lung carcinoma and P388 murine leukemia cells with IC_{50} values of 4.7 and 1.5 μM, respectively (GUNASEKERA et al. 1996). At this time it is not possible to determine whether the growth-inhibitory effects of this molecule may be ascribed to the cdc25-inhibitory activity of this molecule.

2.3 Antisense Inhibitors of CDKs

Antisense technology is an area which has been filled with potential and fraught with difficulty, particularly concerning delivery of the antisense molecule to living cells (STEIN and CHENG 1993; WAGNER 1994). There are several examples where antisense DNA have been applied to target CDK activity, principally by targeting the cyclin moiety (MINSHULL et al. 1989, 1991; WALKER and MALLER 1991). These examples have succeeded because they used relatively simple, easily manipulable in vitro systems for their study. Only recently has it been possible to reliably direct antisense molecules against a target RNA in a living cell. Treatment of cells with a C-5 propyne pyrimidine-modified phosphorothioate antisense molecule directed against cdc2 mRNA, delivered by encapsulation in cationic lipids, resulted in the specific downregulation of the cdc2 protein (FLANAGAN et al. 1996). In normal cell types, such as CV-1 cells, this loss of cdc2 protein manifests itself as a cell cycle block in G_2 phase, as would be expected for inhibition of cdc2 activity. No apparent phenotype was observed upon treatment of tumor cell lines. These results are somewhat different from those observed using small-molecule CDK inhibitors (KAUR et al. 1992; WORLAND et al. 1993; GLAB et al. 1994; VESELY et al. 1994; ABRAHAM et al. 1995), where a G_2 phase block may be observed in both normal and tumor cell lines. There are differences between the two approaches which may account for these discrepancies. The antisense approach is highly specific and other non-target CDK protein levels are not directly affected by the cdc2 antisense. In contrast, all of the small-molecule CDK inhibitors identified to date are less specific and target several CDK activities. It is possible, therefore, that in tumor cells, the loss of cdc2 protein may be compensated for by another CDK, such as CDK2. It is also possible that tumor cells are better able to progress through the cell cycle with the small residual cdc2 protein remaining after antisense treatment. Surprisingly, antisense molecules directed against cyclin B do not show significant alteration of cell cycle progression in any cell type, despite the fact that they cause a significant reduction of protein levels (FLANAGAN et al. 1996). These results indicate that the ability of a somatic cell to accommodate losses in individual cell cycle proteins may be compensated by redundancy in functions of the individual proteins. This therefore suggests that, in order to induce a CDC phenotype in somatic cells, it may be necessary to knock out the activity of more than just one CDK-cyclin complex. Antisense approaches will likely continue to provide a useful tool for the future dissection of the specific events in the cell cycle.

3 CDK Inhibitors as Molecular Tools

Much of our knowledge of the role of CDKs in cell cycle control has been learned from investigating the effects of deliberately altering their activity. CDK-inhibitory compounds therefore show a great potential as tools for the dissection of the molecular pathways in which CDKs are involved. Caution should be used in both the design and interpretation of the results from experiments using CDK inhibitors, however. Despite their apparent specificity, it is unlikely that the only intracellular target for CDK inhibitors will be CDKs themselves. Such problems have certainly confused the interpretation of results using staurosporine and its analogues; this inhibitor acts on a wide array of kinases, and it is not possible to determine whether its cell cycle-inhibitory effects are a direct result of CDK inhibition, the result of impacting on upstream or downstream pathways, or a combination of all of these (see above). More accurate interpretation of the results from the use of CDK inhibitors may be made in a number of ways: (a) designing more specific inhibitors, (b) using inactive analogues of the inhibitor, such as iso-olomoucine (VESELY et al. 1994) to demonstrate that the effects are limited to the CDK inhibitor alone, and (c) using several structurally unrelated CDK inhibitors or alternative methods of targeting CDK in the same assay. This last method presents perhaps the most convincing way of ensuring that the small molecule is acting through CDK inhibition of in the cell. In this case, it would be expected that the same results would be generated by all of the inhibitors if CDKs were the target of the molecules. Ultimately, it would also be desirable to devise cell-based assays for CDK activity to confirm that the compound indeed inhibits CDK activity in a cellular environment. These assays have yet to be reported.

There are multiple reports which demonstrate that CDK inhibitors influence cell cycle progression, consistent with their proposed mechanism of action. A wide variety of target cells have been shown to be inhibited for cell cycle progression by the more specific CDK inhibitors olomoucine, flavopiridol, and butyrolactone (CARLSON et al. 1996; ABRAHAM et al. 1994; GLAB et al. 1994; LOSIEWICZ et al. 1994; SOMEYA et al. 1994; KITAGAWA et al. 1993; KAUR et al. 1992), indicating the breadth of utility of these inhibitors. Typically, CDK inhibitors induce a cell cycle arrest in G_1 and G_2 phase, but do not show a pronounced S-phase arrest (ABRAHAM et al. 1994; GLAB et al. 1994; LOSIEWICZ et al. 1994; SOMEYA et al. 1994; KAUR et al. 1992). In particular, WORLAND et al. (1993) showed that flavopiridol causes only a cell cycle delay in S phase at concentrations that induce a potent arrest in both G_1 and G_2 phase. It has been reported that CDK activity, particularly cyclin A–CDK2, is required for progression through S phase (CARDOSO et al. 1993); SOBCZAK-THEPOT et al. 1993; D'URSO et al. 1990). These data all suggest, therefore, that concentrations of CDK inhibitors which can strongly arrest G_1/S- and G_2/M-phase progression cannot fully prevent S-phase progression, consistent with the hypothesis that the progression through S phase requires a lower threshold of CDK activity (STERN and NURSE 1996) (or may not even require CDK activity at all; DEGREGORI et al. 1995) than the entry into either S or M phase. It is likely that the use of CDK inhibitors will further uncover the roles CDK play in the control of S-phase progression.

CDK inhibitors have also been used to probe the role of CDKs in mediating the onset of apoptosis. Apoptosis, or cell death, may be induced by a wide variety of signals (EVAN et al. 1995; MARTIN et al. 1994; HEINTZ 1993). There have been a number of conflicting reports showing either a correlation (YAO et al. 1996; LI et al. 1995; DONALDSON et al. 1994; SHI et al. 1994) or a lack of correlation (FREEMAN et al. 1994; NORBURY et al. 1994; OBERHAMMER et al. 1994) of CDK activation in apoptotic processes. The issue has been confused by the fact that a wide variety of cell types and apoptosis inducing signals have been used. It also is not certain, in all cases, whether the activation of CDK activity in apoptosis is a cause or a consequence of the initiation of cell death. A number of recent reports have used CDK inhibitors as tools to dissect this involvement of CDKs in apoptosis. These experiments are beginning to shed some light on this confusing issue.

PARK and coworkers (1996) demonstrated that both flavopiridol and olomoucine suppress neuronal cell death induced by nerve growth factor (NGF) withdrawal in sympathetic neurons and postmitotic, differentiated PC12 cells. These results support the hypothesis that apoptosis due to neurotrophin withdrawal may be due to an inappropriate entry into the cell cycle; prevention of cell cycle entry by CDK inhibitors may prevent the signals which induce cell death. Interestingly, these authors also showed that the same inhibitors are able to induce cell death in the presence of serum in undifferentiated, proliferating PC12 cells. These results suggest that the role of CDK activity in cell death is not constant and may depend on a number of factors, including cell type, state of proliferation of the cell, and apoptosis-inducing signal.

Another example of the induction of an apoptotic signal is the overexpression of c-myc in serum-starved fibroblasts (EVAN et al. 1992). This event has been shown to correlate with an increase in CDK activity (HOANG et al. 1994). The role of CDK inhibitors in myc-induced apoptosis has been carefully examined (RUDOLPH et al. 1996). By using both the CDK-inhibitory proteins (p16, p21, and p27) as well as the purine analogue inhibitor roscovitine, RUDOLPH et al. (1996) showed that the activation of CDK activity was not necessary for myc-induced apoptosis, but was necessary and sufficient for myc-dependent cell cycle progression, as determined by the upregulation of cyclin A, a late G_1/S-phase cyclin. These examples provide compelling examples of how CDK inhibitors may be used to separate fact from fiction in the dissection of the roles of CDK in cellular processes.

As will be described below CDK inhibitors show potential for use as therapeutic agents. With a greater understanding of the precise roles of CDK in the induction of arrest, apoptosis, and other cell fates in different systems, it may be possible to use CDK inhibitors to deliberately manipulate cell fate in a therapeutic context. In particular, the specific induction of apoptosis in tumor cells is a highly desired outcome in the treatment of cancer. Although many cell types, both normal and tumor, show cell cycle arrests in response to CDK inhibitors, the induction of cell death by CDK inhibitors in proliferating tumor cells has been shown by a number of groups (BIBLE and KAUFMANN 1996; PARK et al. 1996; PARKER et al. 1996; SHIBATA et al. 1996). For example, the inhibition of CDK activity by butyrolactone-1 is sufficient to induce apoptosis in proliferating HL60 cells, but not U937 cells (SHIBATA et al. 1996). Cell death in HL60 cells occurs predominantly from the G_1/S phase of the cell cycle, presumably because of aberrant progression of cells into S-phase in the absence of appropriate signals. U937 cells accumulated in G_2 phase, suggesting that, although the G_1-phase checkpoint was bypassed in these cells, the signals inducing apoptosis in HL60 cells are not present in all cell types. We (K. Dold and D.H. Walker, unpublished results) and others (BIBLE and KAUFMANN 1996; PARKER et al. 1996) have confirmed that some cell types respond to inhibition of CDK activity by apoptosis, whereas others respond by arresting. These results suggest that, in some tumor cells, the loss of CDK activity and cell cycle arrest cannot be tolerated. Dissection of the molecular signals which are involved in determining apoptosis or arrest will likely reveal a subset of tumors which will respond well to treatment with CDK inhibitors.

The possibilities for CDK inhibitors as molecular tools are endless. Future experiments using small-molecule CDK inhibitors will undoubtedly advance our understanding of the role these proteins play in determining the fate of cells. Further experiments may investigate the role of CDK in other processes, such as differentiation. These compounds will also provide a means of investigating the role of CDK5 in nervous system disorders such as Alzheimer's disease (HOSOI et al. 1995).

4 CDK Inhibitors as Therapies

Progressing from a compound with desirable activity in a cell-based model to one with good activity in vivo is a difficult step. Very few molecules are able to make this leap. In order to generate a compound with the requisite pharmacologic properties and low toxicity, it is often necessary to identify molecules with a high degree of potency and selectivity. The majority of the currently reported CDK inhibitors have yet to accomplish this goal. Despite these high hurdles, the future for CDK inhibitors as therapies is particularly bright.

As outlined above, one outcome of the use of CDK inhibitors as molecular tools is the projection of potential ways in which these compounds may be used as therapeutics. Of particular interest is the potential use of CDK inhibitors in the treatment of cancer and other hyperproliferative disorders. To be of any real therapeutic benefit, it will be necessary to demonstrate that CDK-inhibitory compounds show an improved therapeutic outcome when compared to currently approved therapies. Some of the experimental approaches outlined above have suggested that CDK inhibitors may show stand-alone antitumor activity in certain cases, both as cytostatics (KAUR et al. 1994) and as cytotoxics (PARK et al. 1996; PARKER et al. 1996; SHIBATA et al. 1996). It is also possible that the specific inhibition of cdc2 may synergistically increase the cytotoxic activity of some chemotherapeutic agents in tumor cells. Indeed, it has been shown that CDK inhibitors may increase the antitumor activity of cytotoxics (ONGKEKO et al. 1995; SCHWARTZ et al. 1996), possibly by inducing tumor cells to arrest or delay in phases of the cell cycle in which they are more sensitive to the cytotoxic. FLANAGAN et al. (1996) additionally showed that cdc2 antisense molecules were able to induce a cell cycle arrest in G_2 phase in normal, but not tumor cells. This differential response of normal and tumor cells may be exploitable as a way to protect normal cells from agents which work in mitosis. Protecting cells from cytotoxic activity by preventing cell cycle progression has been hypothesized previously (KOHN et al. 1994) and has been demonstrated, either with CDK inhibitors (STONE et al. 1996) or with cell cycle-inhibitory compounds (TSAO et al. 1992). This concept is further supported by the observation that taxol-induced apoptosis appears to require cell cycle progression (DONALDSON et al. 1994). Ideally, it would be desirable to identify conditions in which a CDK-inhibitory compound would capture elements of all of these potential activities, i.e., an antitumor activity which synergizes with current cytotoxic therapy, while protecting from the toxic side effects of those cytotoxics. Such an approach could radically alter our approach and viewpoint towards the treatment of cancer.

The only reported CDK inhibitor which is under clinical evaluation at this time is flavopiridol (E. Sausville and A. Senderowicz, personal communication). In preclinical evaluation, flavopiridol was shown to have activity against a number of xenografted tumor cell lines. CZECH et al. (1995) reported that flavopiridol inhibited the growth of 18 out of 22 xenografted tumors. In some cases, the activity of flavopiridol was comparable to that of commonly used cytotoxic drugs. This activity appeared to be primarily cytostatic, since the compound was unable to induce tumor regression in any of the models used. Flavopiridol is currently being evaluated in phase-1 clinical trials, (E. Sausville and A. Senderowicz, personal communication). The initial results show that flavopiridol is well tolerated, with diarrhea as the dose-limiting toxicity. Hematopoetic toxicity, which is common with cytotoxic anticancer drugs, has not been found to be a dose-limiting side effect. It is difficult to evaluate the therapeutic potential of flavopiridol at this early stage. Several patients have shown no disease progression for up to 1 year, however, consistent with a proposed cytostatic mechanism for this drug. Given that flavopiridol is generally well tolerated, it shows promise as a long-term therapy to stabilize disease.

Although flavopiridol is the most potent CDK inhibitor identified to date (CARLSON et al. 1996; LOSIEWICZ et al. 1994) and shows some promise in the clinic (E. Sausville and A. Senderowicz, personal communication), it certainly does not represent the pinnacle of possibilities for a CDK inhibitor. Identification of more potent, selective inhibitors of CDK4 or cdc2–CDK2 is the goal of a number of pharmaceutical and biotechnology companies as well as several academic laboratories. These efforts, along with the identification of better ways to use these molecules as a result of cell-based experimental approaches, will hopefully build on these initial successes. It is likely that we will see several novel potent and selective CDK inhibitors progress to the clinic within the next decade.

The use of these molecules to treat other disorders which result from inappropriate proliferation will likely follow closely behind. These may include inflammatory diseases (such as psoriasis), restenosis, and viral infections. The current small molecular weight inhibitors of CDK activity may provide the tools to identify and validate these treatments.

Acknowledgements. I would like to thank Dr. Laurent Meijer and Dr. Ed Sausville for kindly sharing unpublished data. I would also like to thank Lee Kuyper, Brett Lovejoy, and Stephen Davis for input and critical reading of this manuscript.

References

Abraham RT, Acquarone M, Anderson A, Asensi A, Belle R, Berger F, Bergounioux C, Brunn G, Buquet-Fagot C, Fagot D, Glab N, Goudeau H, Goudeau M, Guerrier P, Houghton P, Hendricks H, Kloareg B, Lippai M, Marie D, Maro B, Meijer L, Mester J, Mulner-Lorillon O, Poulet SA, Scheirenberg E, Schutte B, Vaulot D, Verlhac MH (1994) Cellular effects of olomoucine, an inhibitor of cyclin-dependent kinases, Biol Cell 83:105–120

Akinaga S, Nomura K, Gomi K, Okabe M (1994) Effect of UCN-01, a selective inhibitor of protein kinase C, on the cell-cycle distribution of human epidermoid carcinoma, A431 cells. Cancer Chemother Pharmacol 33:273–280

Almasan A, Linke SP, Paulson TG, Huang LC, Wahl GM (1995) Genetic instability as a consequence of inappropriate entry into and progression through S-phase. Cancer Metastasis Rev 14:59–73

Azevedo WF, Mueller-Dieckmann HJ, Schulfe-Gahmen U, Worland PJ, Sausville E, Kim SH (1996) Structural basis for specificity and potency of a flavinid inhibitor of human CDK2, a cell cycle kinase. Proc Natl Acad Sci USA 93:2735–2740

Bible KC, Kaufmann SH (1996) Flavopiridol: a cytotoxic flavone that induces cell death in noncycling A549 human lung carcinoma cells. Cancer Res 56:4856–4861

Bojanowski K, Nishio K, Fukuda M, Larsen AK, Saijo N (1994) Effect of suramin on p34^{cdc2} kinase in vitro and in extracts from human H69 cells: evidence for a double mechanism of action. Biochem Biophys Res Commun 203:1574–1580

Cardoso MC, Leonhardt H, Nadal-Ginard B (1993) Reversal of terminal differentiation and control of DNA replication: cyclin A and CDK2 specifically localize at subnuclear sites of DNA replication. Cell 74:979–992

Carlson BA, Dubay MM, Sausville EA, Brizuela L, Worland PJ (1996) Flavopiridol induces G1 arrest with inhibition of cyclin-dependent kinase (CDK) 2 and CDK4 in breast carcinoma cells. Cancer Res 56:2973–2978

Chen J, Jackson PK, Kirschner MW, Dutta A (1995) Separate domains of p21 involved in the inhibition of cdk kinase and PCNA. Nature 374:386–388

Colas P, Cohen B, Jenssen T, Grishina I, McCoy J, Brent R (1996) Genetic selection of peptide aptamers that recognize and inhibit cyclin-dependent kinase 2. Nature 380:548–550

Czech J, Hoffmann D, Naik R, Sedlacek H-H (1995) Antitumoral activity of flavone L86-8275. Int J Oncol 6:31–36

DeGregori J, Leone G, Ohtani K, Miron A, Nevins JR (1995) E2F-1 accumulation bypasses a G1 arrest resulting from the inhibition of G1 cyclin-dependent kinase activity. Genes Dev 9:2873–2878

Donaldson KL, Goolsby GL, Kiener PA, Wahl AF (1994) Activation of p34^{cdc2} coincident with taxol-induced apoptosis. Cell Growth Differ 5:1041–1050

D'Urso G, Marraccino RL, Marshak DR, Roberts JM (1990) Cell cycle control of DNA replication by a homologue from human cells of the p34^{cdc2} protein kinase. Science 250:786–791

Evan GI, Wyllie AH, Gilbert CS, Littlewood TD, Land H, Brooks M, Waters CM, Penn LZ, Hancock DC (1992) Induction of apoptosis in fibroblasts by c-myc protein. Cell 69:119–128

Evan GI, Brown L, Whyte M, Harington E (1995) Apoptosis and the cell cycle. Curr Opin Cell Biol 7:825–834

Fahraeus R, Paramio JM, Ball KL, Lain S, Lane DP (1996) Inhibition of pRb phosphorylation and cell-cycle progression by a 20-residue peptide derived from p16$^{CDKN2/INK4A}$. Curr Biol 6:84–91

Flanagan WM, Su LL, Wagner RW (1996) Elucidation of gene function using C-5 propyne antisense oligonucleotides. Nature Biotechnol 14:1139–1145

Freeman RS, Estus S, Johnson EM (1994) Analysis of cell-cycle-related gene-expression in postmitotic neurons – selective induction of cyclin D1 during programmed cell death. Neuron 12:343–355

Gadbois DM, Hamaguchi JR, Swank RW, Bradbury EM (1992) Staurosporine is a potent inhibitor of p34^{cdc2} and p34^{cdc2}-like kinases. Biochem Biophys Res Commun 184:80–85

Galaktionov K, Lee AK, Eckstein J, Draetta G, Meckler J, Loda M, Beach D (1995) CDC25 phosphatases as potential human oncogenes. Science 269:1575–1577

Glab N, Labidi B, Qin L-X, Trehin C, Bergounioux C, Meijer L (1994) Olomoucine, an inhibitor of the cdc2/cdk2 kinases activity, blocks plant cells at the G1 to S and G2 to M cell cycle transitions. FEBS Lett 353:207–211

Gunasekera SP, McCarthy PJ, Kelly-Borges M, Lobkovsky E, Clardy J (1996) Dysidiolide: a novel protein phosphatase inhibitor from the Caribbean sponge *Dysidea etheria* de Laubenfels. J Am Chem Soc 118:8759–8760

Heintz N (1993) Cell death and the cell cycle: a relationship between transformation and neurodegeneration? Trends Biochem Sci 18:157–159

Hoang AT, Cohen KJ, Barrett JF, Bergstrom DA, Dang CV (1994) Participation of cyclin A in myc-induced apoptosis. Proc Natl Acad Sci USA 91:6875–6879

Horiguchi T, Nishi K, Hakoda S, Tanida S, Nagata A, Okayama H (1994) DnacinA1 and DnacinB1 are antitumour antibiotics that inhibit cdc25B phosphatase activity. Biochemical Pharmacol 48:2139–2141

Hosoi T, Uchiyama M, Okumura E, Saito T, Uchida T, Okuyama A, Kishimoto T, Hisanaga S (1995) Evidence for cdk5 as a major activity phosphorylating tau protein in porcine brain extract. J Biochem 117:741–749

Hunter T (1995) Protein kinases and phosphatases: the yin and yang of protein phosphorylation and signaling. Cell 80:225–236

Jeffrey PD, Russo AA, Polyak K, Gibbs E, Hurwitz J, Massague J, Pavletich NP (1995) Mechanism of CDK activation revealed by the structure of a cyclinA-CDK2 complex. Nature 376:313–320

Kaur G, Stetler-Stevenson M, Sebers S, Worland P, Sedlacek H, Myers C, Czesh J, Naik R, Sausville E (1992) Growth inhibition with reversible cell cycle arrest of carcinoma cells by flavone L86-8275. J Natl Cancer Inst 84:1736–1740

Kawakami K, Futami H, Takahara J, Yamaguchi K (1996) UCN-01, 7-hydroxyl-staurosporine, inhibits kinase activity of cyclin-dependent kinases and reduces the phosphorylation of the retinoblastoma susceptibility gene product in A549 human lung cancer cell line. Biochem Biophys Res Commun 219:778–783

Kitagawa M, Okabe T, Ogino H, Matsumoto H, Suzuki-Takahasi I, Kokubo T, Higashi H, Saitoh S, Taya Y, Yasuda H, Ohba Y, Nishimura S, Tanaka N, Okuyama A (1993) Butyrolactone I, a selective inhibitor of cdk2 and cdc2 kinase. Oncogene 8:2425–2432

Kohn KW, Jackman J, O'Connor PM (1994) Cell cycle control and cancer chemotherapy. J Cell Biochem 54:440–452

Larsen AK (1993) Suramin: an anticancer drug with unique biological effects. Cancer Chemother Pharmacol 32:96–98

Lees E (1995) Cyclin dependent kinase regulation. Curr Opin Cell Biol 7:773–780

Li CJ, Friedman DJ, Wang C, Metlev V, Pardee A (1995) Induction of apoptosis in uninfected lymphocytes by HIV-1 TAT protein. Science 268:429–431

Losiewicz MD, Carlson BA, Kaur GK, Sausville EA, Worland PJ (1994) Potent inhibition of cdc2 kinase activity by the flavinoid L86-8275. Biochem Biophys Res Commun 201:589–595

Martin SJ, Green DR, Cotter TG (1994) Dicing with death: dissecting the components of the apoptosis machinery. Trends Biochem Sci 19:26–30

Meijer L (1995) Chemical inhibitors of cyclin-dependent kinases. In: Meijer L, Guidet S, Tung HYL (eds) Progress in cell cycle research, vol I. Plenum, New York, pp 351–363

Meijer L (1996) Chemical inhibitors of cyclin dependent kinases. Trends Cell Biol 6:384–387

Minshull J, Blow JJ, Hunt T (1989) Translation of cyclin mRNA is necessary for extracts of activated Xenopus eggs to enter mitosis. Cell 56:947–956

Minshull J, Murray A, Colman A, Hunt T (1991) Xenopus oocyte maturation does not require new cyclin synthesis. J Cell Biol 114:767–772

Morgan DO (1995) Principles of CDK regulation. Nature 374:131–134

Norbury C, MacFarlane M, Fearnhead H, Cohen GM (1994) cdc2 activation is not required for thymocyte apoptosis. Biochem Biophys Res Commun 202:1400–1406

Oberhammer FA, Hochegger K, Froschl G (1994) Chromatin condensation during apoptosis is accompanied by degradation of lamin A and B without enhanced activation of cdc2 kinase. J Cell Biol 126:827–837

Ogawara H, Akiyama T, Watanabe S, Ito N, Kobori M, Seoda Y (1989) Inhibition of tyrosine protein kinase activity by synthetic isoflavones and flavones. J Antibiot 42:340–343

Ohashi M, Sugikawa E, Nakanishi N (1995) Inhibition of p53 protein phosphorylation by 9-hydroxyellipticine: a possible anticancer mechanism. Jpn J Cancer Res 86:819–825

Omura S, Iwai Y, Hirao A, Nakagawa A, Awaya J, Tsuchiya H, Takahashi Y, Masuma R (1977) A new alkaloid AM-2282 of Streptomyces origin. Taxonomy, fermentation, isolation and preliminary characterization. J Antibiot 30:275–281

Ongkeko W, Ferguson DJP, Harris AL, Norbury C (1994) Inactivation of cdc2 increases the level of apoptosis induced by DNA damage. J Cell Sci 108:2897–2904

Park DS, Farinelli SE, Greene LA (1996) Inhibitors of cyclin-dependent kinases promote survival of postmitotic neuronally differentiated PC12 cells and sympathetic neurons. J Biol Chem 271:8161–8169

Parker BW, Senderowicz AM, Nieves-Neira W, Pommier Y, Sausville EA (1996) DNA fragmentation and apoptosis of lymphoma and prostate cancer cell lines after flavopiridol treatment. Proc Am Cancer Res 37:398

Poon RYC, Hunter T (1995) Innocent bystanders or chosen collaborators? Curr Biol 5:1243–1247

Rialet V, Meijer L (1991) A new screening test for antimitotic compounds using the universal M-phase-specific protein kinase, $p34^{cdc2}$/cyclinB^{cdc13}, affinity-immobilized on $p13^{suc1}$-coated microtitration plates. Anticancer Res 11:1581–1590

Rudolph B, Saffrich R, Zwicker J, Henglein B, Muller R, Ansorge W, Eilers M (1996) Activation of cyclin-dependent kinases by Myc mediates induction of cyclin A, but not apoptosis. EMBO J 15:3065–3076

Russo AA, Jeffery PD, Patten AK, Massague J, Pavletich NP (1996) Crystal structure of the $p27^{Kip1}$ cyclin dependent kinase inhibitor bound to the cyclin A-cdk2 complex. Nature 382:325–331

Schnier JB, Gadbois DM, Nishi K, Bradbury EM (1994) The kinase inhibitor staurosporine induces G1 arrest at two points: effect on retinoblastoma protein phosphorylation and cyclin-dependent kinase 2 in normal and transformed cells. Cancer Res 54:5959–5963

Schulze-Gahmen U, Brandsen J, Jones HD, Morgan DO, Meijer L, Vesely J, Kin SH (1995) Multiple modes of ligand recognition: crystal structures of cyclin-dependent protein kinase 2 in complex with ATP and two inhibitors, olomoucine and isopentenyladenine. Proteins 22:378–391

Schwartz GK, Farsi K, Danso D, Dhupar SK, Kelsen D, Spriggs D (1996) The protein kinase C (PKC) inhibitors UCN-01 and flavopiridol (FLAVO) significantly enhance the cytotoxic effect of chemotherapy by promoting apoptosis in gastric and breast cancer cells. Proc Am Soc Clin Oncol 15:207

Seynaeve CM, Stetler-Stevenson M, Sebers S, Kaur GA, Sausville EA, Worland PJ (1993) Cell cycle arrest and growth inhibition by the protein kinase antagonist UCN-01 in human breast carcinoma cells. Cancer Res 53:2081–2086

Sherr CJ, Roberts JM (1995) Inhibitors of mammalian G1 cyclin-dependent kinases. Genes Dev 9:1149–1163

Shi LS, Nishioika WK, Th'ing J, Bradbury EM, Lichfield DW, Greenberg AH (1994) Premature p34^{cdc2} activation required for apoptosis. Science 263:1143–1145

Shibata Y, Nishimura S, Okuyama A, Nakamura T (1996) p53-independent induction of apoptosis by cyclin-dependent kinase inhibition. Cell Growth Differ 7:887–891

Sobczak-Thepot J, Harper F, Florentin Y, Zindy F, Brechot C, Puvion E (1993) Localization of cyclin A at the sites of cellular DNA replication. Exp Cell Res 206:43–48

Someya A, Tanaka N, Okuyama A (1994) Inhibition of cell cycle oscillation of DNA replication by a selective inhibitor of the cdc2 kinase family, butyrolactone I, in Xenopus egg extracts. Biochem Biophys Res Commun 198:536–545

Stein CA, Cheng Y-C (1993) Antisense oligonucleotides as therapeutic agents – is the bullet really magical? Science 261:1004–1012

Stern B, Nurse P (1996) A quantitative model for the cdc2 control of S phase and mitosis in fission yeast. Trends Genet 12:345–350

Stone S, Dayananth P, Kamb A (1996) Reversible, p16-mediated cell cycle arrest as protection from chemotherapy. Cancer Res 56:3199–3202

Takahashi I, Saitoh Y, Yoshida M, Sano H, Nakano H, Morimoto M, Tamaoki T (1989) UCN-01 and UCN-02, new selective inhibitors of protein kinase C. II. Purification, physico-chemical properties, structural determination and biological activities. J Antibiot 42:571–576

Tsao Y-P, D'Arpa P, Liu LF (1992) The involvement of active DNA synthesis in camptothecin-induced G2 arrest: altered regulation of p34^{cdc2}/cyclinB. Cancer Res 52:1823–1829

Vesely J, Havlicek L, Strnad M, Blow JJ, Donella-Deana A, Pinna L, Letham DS, Kato J-Y, Detivaud L, LeClerc S, Meijer L (1994) Inhibition of cyclin-dependent kinases by purine analogues. Eur J Biochem 224:771–786

Wagner RW (1994) Gene inhibition using antisense oligodeoxynucleotides. Nature 372:333–335

Walker DH, Maller JL (1991) Role for cyclin A in the dependence of mitosis on completion of DNA replication. Nature 354:314–317

Wang Q, Worland PJ, Clark JL, Carlson BA, Sausville EA (1995) Apoptosis in 7-hydroxystaurosporine-treated T lymphoblasts correlates with activation of cyclin-dependent kinases 1 and 2. Cell Growth Differ 6:927–936

Worland PJ, Kaur G, Stetler-Stevenson M, Sebers S, Sartor O, Sausville EA (1993) Alteration of the phosphorylation state of p34^{cdc2} kinase by the flavone L86-8275 in breast carcinoma cells. Biochem Pharmacol 46:1831–1840

Yao SL, McKenna KA, Sharkis SJ, Bedi A (1996) Requirement of p34^{cdc2} kinase for apoptosis mediated by the Fas/Apo-1 receptor and interleukin 1β-converting enzyme-related proteases. Cancer Res 56:4551–4555

Subject Index

A
adenovirus E1A protein 32
AMP, cyclic (cAMP) 80
ankyrin repeat 3
antisense inhibitors 157, 158
anucleate cells 18
apoptosis 129, 159
asymmetric cell division 10
ATP 152

B
B-type cyclins 6
Beckwith-Weidemann syndrome 35
BRCA1 144
bud emergence 15
butyrolactone 1 150, 154

C
cAMP (see AMP, cyclic)
cancer 139–146, 161
Cdc5 7
CDC5 7
Cdc13 17
CDC15 7
Cdc18 18
cdc25 phosphatase 157
cdc34, ubiquitination 13
CDK
– antisense inhibitors 157, 158
– chemical inhibitors 149–155
– inhibitors 107, 139
– – expression 143
– – putative physiological role 117, 118
– peptide inhibitors 156
– substrate, SICI 6
Cdk-cyclin-p27 complex, X-ray crystallographic analysis 31
CDK2 109
cell cycle 59
– progression 60, 61
– proteins 60, 61
chemical inhibitors
– butyrolactone-1 154

– flavonoids 153
– purine analogues 151
– staurosporine analogues 154
chromosomal
– integrity 10
– rearrangements 10
– segregation 18
chromosome
– breakage 10
– loss 10
Cig1 17
Cig2 17
Cip/Kip gene family 107
Clb2 cyclin, destruction 7
CLN3 5
contact inhibition 32
corpus luteum 114
CSF-1 80
cut phenotype 18
cyclin D-CDK 109
cyclin D_2 114
cyclin E 114

D
D/CDK4 44
daughter cells 10
dbf2 7
DNA replication 6, 9
– overreplication 17–19
DNA-damage checkpoint 27

E
embryonic fibroblasts 108

F
Far1 16
– proteolysis 13
– transcription 11
– ubiquitination 13
FAR1, transcription 13
flavonoids 153
flavopiridol 150
Fus3 14

G

G_1 cyclins, inhibition 11
G_1-phase 10, 27, 111
– Far1 16
– SICI involvement in 10
G_2/M phase 10
– SICI involvement in 10
genome stability 27
growth hormone (GH) 111

H

hematopoietic cell differentiation 68–71
HL-525 64
HL-60 64, 70
homozygous deletion 140
human cancer 139–146, 161

I

IFN-β 64–68, 73
IGF (see insulin-like growth factor)
IL-2 81
IL-6 69
in situ hybridization 107
INK4 family 43–52
insulin-like growth factor (IGF) 111
– IGF II 111
ionizing radiation 26

K

keratinocyte differentiation 72
KSS1 14

L

LIM domains 16
– Far1 16

M

MAP kinase 11, 14
– Fus3 14
– KSS1 14
mating
– S. pombe 19
– yeast 10
medial nuclear division 10
melanoma 141
– differentiation 64–68
Mem1, transcription 13
mitogen-activated kinase, Fus3 11
MLM 141, 142
mouse embryonic stem cells 105
muscle cell differentiation 71, 72
MyoD 71, 72
myrercetin 150

N

N^6-isopentenyladenine 150
nitrogen starvation 19
nocodazole 13

O

olomucine 150
oriented growth 15
ovary 112
overreplication 17–19

P

p15^{INK4b} 3, 49, 50, 88
p16 90, 107, 115–117, 142, 145
– and cellular differentiation 90
– role in development 90
p16-null mice, tumor susceptibility 91
p16$^{-/-}$ mouse embryo fibroblasts, properties 91
p16^{INK4a} 3, 45–49
p18INK4c 50–52, 88, 92
p19INK4d 50–52, 88, 92
p21 63–76, 107, 108, 130–132
– cellular differentiation 64
– cloning 63
– development 73, 74
– function 122–125
– – stoichiometry 29
– identification 63
– mode of action 74–76
– regulation 125–127
– structure 122
p21$^{Cip1/Waf1/Sdi1}$ 26–30
p27 107, 109
– cell cycle 33
– degradation 33
– differentiation and proliferation 79–82
– levels in cells, regulation 79
– mechanism of cyclin-dependent kinase inhibition 86
– translation 33
p27-null mice
– abnormalities of the thymus 83
– phenotype 83
p27$^{-/-}$ mice
– growth arrest in fibroblasts 86
– pituitary hyperplasia 84
– reproductive system abnormalities 85
– retinal structure 85
p27^{Kip1} 30–34, 76–79
p53 26, 127–132
p57 107
p57^{Kip2} 34, 35, 86, 87
PCNA (see proliferating cell nuclear antigen)
peptide inhibitors 157
pheromone 12
Pho4, phosphorylation 4

PHO4 3
PHO5, transcription 5
Pho80 3
Pho81 3
– transcription 4
phosphatases 2
– transcription 5
phosphorylation
– Pho4 4
– SIC1 6
Plk1 7
polo 7
positive feedback loop 4
praafian fillicles 112
proliferating cell nuclear antigen (PCNA) 29, 35, 74, 75, 132
protein kinase
proteolysis
– CLB 7
– Far1 13
– SIC1 8
Puc1 17
purine analogues 151

R
radiation
– ionizing 26
– ultraviolet 26
rapamycin 32, 33, 111
retina 114
roscovitine 150

S
S phase initiation 6, 9
senescence 28
SIC1 6
– phosphorylation 6
– proteolysis 8

– transcription 7
– ubiquitination 8
simian virus 40 (SV40) 75
staurosporine 150
– analogues 154
Ste12, transcription 13
SV40 (simian virus 40) 75

T
T cell receptor (TCR)-β chain 108
TGF-β 32, 33, 72, 77, 81
TPA 70
transcription
– *cdc13* 19
– FAR1 11
– *FAR1* 13
– Mem1 13
– *PHO5* 5
– Pho81 4
– phosphatase 5
– *rum1* 19
– Ste12 13
transforming growth factor (TGF)-β 111
tumor suppressor 109
tumorigenesis 140

U
ubiquitination
– *CDC4* 9
– *CDC34* 8
– *cdc34* 13
– *CDC53* 9
– Far1 13
– SIC1 8
ultraviolet radiation 26

X
Xenopus Kip proteins 35

Printing: Saladruck, Berlin
Binding: Buchbinderei Lüderitz & Bauer, Berlin

Current Topics in Microbiology and Immunology

Volumes published since 1989 (and still available)

Vol. 187: **Rupprecht, Charles E.; Dietzschold, Bernhard; Koprowski, Hilary (Eds.):** Lyssaviruses. 1994. 50 figs. IX, 352 pp. ISBN 3-540-57194-9

Vol. 188: **Letvin, Norman L.; Desrosiers, Ronald C. (Eds.):** Simian Immunodeficiency Virus. 1994. 37 figs. X, 240 pp. ISBN 3-540-57274-0

Vol. 189: **Oldstone, Michael B. A. (Ed.):** Cytotoxic T-Lymphocytes in Human Viral and Malaria Infections. 1994. 37 figs. IX, 210 pp. ISBN 3-540-57259-7

Vol. 190: **Koprowski, Hilary; Lipkin, W. Ian (Eds.):** Borna Disease. 1995. 33 figs. IX, 134 pp. ISBN 3-540-57388-7

Vol. 191: **ter Meulen, Volker; Billeter, Martin A. (Eds.):** Measles Virus. 1995. 23 figs. IX, 196 pp. ISBN 3-540-57389-5

Vol. 192: **Dangl, Jeffrey L. (Ed.):** Bacterial Pathogenesis of Plants and Animals. 1994. 41 figs. IX, 343 pp. ISBN 3-540-57391-7

Vol. 193: **Chen, Irvin S. Y.; Koprowski, Hilary; Srinivasan, Alagarsamy; Vogt, Peter K. (Eds.):** Transacting Functions of Human Retroviruses. 1995. 49 figs. IX, 240 pp. ISBN 3-540-57901-X

Vol. 194: **Potter, Michael; Melchers, Fritz (Eds.):** Mechanisms in B-cell Neoplasia. 1995. 152 figs. XXV, 458 pp. ISBN 3-540-58447-1

Vol. 195: **Montecucco, Cesare (Ed.):** Clostridial Neurotoxins. 1995. 28 figs. XI., 278 pp. ISBN 3-540-58452-8

Vol. 196: **Koprowski, Hilary; Maeda, Hiroshi (Eds.):** The Role of Nitric Oxide in Physiology and Pathophysiology. 1995. 21 figs. IX, 90 pp. ISBN 3-540-58214-2

Vol. 197: **Meyer, Peter (Ed.):** Gene Silencing in Higher Plants and Related Phenomena in Other Eukaryotes. 1995. 17 figs. IX, 232 pp. ISBN 3-540-58236-3

Vol. 198: **Griffiths, Gillian M.; Tschopp, Jürg (Eds.):** Pathways for Cytolysis. 1995. 45 figs. IX, 224 pp. ISBN 3-540-58725-X

Vol. 199/I: **Doerfler, Walter; Böhm, Petra (Eds.):** The Molecular Repertoire of Adenoviruses I. 1995. 51 figs. XIII, 280 pp. ISBN 3-540-58828-0

Vol. 199/II: **Doerfler, Walter; Böhm, Petra (Eds.):** The Molecular Repertoire of Adenoviruses II. 1995. 36 figs. XIII, 278 pp. ISBN 3-540-58829-9

Vol. 199/III: **Doerfler, Walter; Böhm, Petra (Eds.):** The Molecular Repertoire of Adenoviruses III. 1995. 51 figs. XIII, 310 pp. ISBN 3-540-58987-2

Vol. 200: **Kroemer, Guido; Martinez-A., Carlos (Eds.):** Apoptosis in Immunology. 1995. 14 figs. XI, 242 pp. ISBN 3-540-58756-X

Vol. 201: **Kosco-Vilbois, Marie H. (Ed.):** An Antigen Depository of the Immune System: Follicular Dendritic Cells. 1995. 39 figs. IX, 209 pp. ISBN 3-540-59013-7

Vol. 202: **Oldstone, Michael B. A.; Vitković, Ljubiša (Eds.):** HIV and Dementia. 1995. 40 figs. XIII, 279 pp. ISBN 3-540-59117-6

Vol. 203: **Sarnow, Peter (Ed.):** Cap-Independent Translation. 1995. 31 figs. XI, 183 pp. ISBN 3-540-59121-4

Vol. 204: **Saedler, Heinz; Gierl, Alfons (Eds.):** Transposable Elements. 1995. 42 figs. IX, 234 pp. ISBN 3-540-59342-X

Vol. 205: **Littman, Dan R. (Ed.):** The CD4 Molecule. 1995. 29 figs. XIII, 182 pp. ISBN 3-540-59344-6

Vol. 206: **Chisari, Francis V.; Oldstone, Michael B. A. (Eds.):** Transgenic Models of Human Viral and Immunological Disease. 1995. 53 figs. XI, 345 pp. ISBN 3-540-59341-1

Vol. 207: **Prusiner, Stanley B. (Ed.):** Prions Prions Prions. 1995. 42 figs. VII, 163 pp. ISBN 3-540-59343-8

Vol. 208: **Farnham, Peggy J. (Ed.):** Transcriptional Control of Cell Growth. 1995. 17 figs. IX, 141 pp. ISBN 3-540-60113-9

Vol. 209: **Miller, Virginia L. (Ed.):** Bacterial Invasiveness. 1996. 16 figs. IX, 115 pp. ISBN 3-540-60065-5

Vol. 210: **Potter, Michael; Rose, Noel R. (Eds.):** Immunology of Silicones. 1996. 136 figs. XX, 430 pp. ISBN 3-540-60272-0

Vol. 211: **Wolff, Linda; Perkins, Archibald S. (Eds.):** Molecular Aspects of Myeloid Stem Cell Development. 1996. 98 figs. XIV, 298 pp. ISBN 3-540-60414-6

Vol. 212: **Vainio, Olli; Imhof, Beat A. (Eds.):** Immunology and Developmental Biology of the Chicken. 1996. 43 figs. IX, 281 pp. ISBN 3-540-60585-1

Vol. 213/I: **Günthert, Ursula; Birchmeier, Walter (Eds.):** Attempts to Understand Metastasis Formation I. 1996. 35 figs. XV, 293 pp. ISBN 3-540-60680-7

Vol. 213/II: **Günthert, Ursula; Birchmeier, Walter (Eds.):** Attempts to Understand Metastasis Formation II. 1996. 33 figs. XV, 288 pp. ISBN 3-540-60681-5

Vol. 213/III: **Günthert, Ursula; Schlag, Peter M.; Birchmeier, Walter (Eds.):** Attempts to Understand Metastasis Formation III. 1996. 14 figs. XV, 262 pp. ISBN 3-540-60682-3

Vol. 214: **Kräusslich, Hans-Georg (Ed.):** Morphogenesis and Maturation of Retroviruses. 1996. 34 figs. XI, 344 pp. ISBN 3-540-60928-8

Vol. 215: **Shinnick, Thomas M. (Ed.):** Tuberculosis. 1996. 46 figs. XI, 307 pp. ISBN 3-540-60985-7

Vol. 216: **Rietschel, Ernst Th.; Wagner, Hermann (Eds.):** Pathology of Septic Shock. 1996. 34 figs. X, 321 pp. ISBN 3-540-61026-X

Vol. 217: **Jessberger, Rolf; Lieber, Michael R. (Eds.):** Molecular Analysis of DNA Rearrangements in the Immune System. 1996. 43 figs. IX, 224 pp. ISBN 3-540-61037-5

Vol. 218: **Berns, Kenneth I.; Giraud, Catherine (Eds.):** Adeno-Associated Virus (AAV) Vectors in Gene Therapy. 1996. 38 figs. IX, 173 pp. ISBN 3-540-61076-6

Vol. 219: **Gross, Uwe (Ed.):** Toxoplasma gondii. 1996. 31 figs. XI, 274 pp. ISBN 3-540-61300-5

Vol. 220: **Rauscher, Frank J. III; Vogt, Peter K. (Eds.):** Chromosomal Translocations and Oncogenic Transcription Factors. 1997. 28 figs. XI, 166 pp. ISBN 3-540-61402-8

Vol. 221: **Kastan, Michael B. (Ed.):** Genetic Instability and Tumorigenesis. 1997. 12 figs. VII, 180 pp. ISBN 3-540-61518-0

Vol. 222: **Olding, Lars B. (Ed.):** Reproductive Immunology. 1997. 17 figs. XII, 219 pp. ISBN 3-540-61888-0

Vol. 223: **Tracy, S.; Chapman, N. M.; Mahy, B. W. J. (Eds.):** The Coxsackie B Viruses. 1997. 37 figs. VIII, 336 pp. ISBN 3-540-62390-6

Vol. 224: **Potter, Michael; Melchers, Fritz (Eds.):** C-Myc in B-Cell Neoplasia. 1997. 94 figs. XII, 291 pp. ISBN 3-540-62892-4

Vol. 225: **Vogt, Peter K.; Mahan, Michael J. (Eds.):** Bacterial Infection: Close Encounters at the Host Pathogen Interface. 1998. 15 figs. IX, 169 pp. ISBN 3-540-63260-3

Vol. 226: **Koprowski, Hilary; Weiner, David B. (Eds.):** DNA Vaccination/Genetic Vaccination. 1998. 31 figs. XVIII, 198 pp. ISBN 3-540-63392-8